Trainee Guide

Core Curricula

National Center for Construction Education and Research

Wheels of Learning
Standardized Craft Training

Prentice Hall

Upper Saddle River, New Jersey   Columbus, Ohio

This volume is one of many in the *Wheels of Learning* craft training program. This program, covering more than 20 standardized craft areas, including all major construction skills, was developed over a period of years by industry and education specialists. Sixteen of the largest construction and maintenance firms in the U.S. committed financial and human resources to the teams that wrote the curricula and planned the national accredited training process. These materials are industry-proven and consist of competency-based textbooks and instructor guides.

Wheels of Learning was developed by the National Center for Construction Education and Research in response to the training needs of the construction and maintenance industries. The NCCER is a nonprofit educational entity supported by the following industry and craft associations:

- Associated Builders and Contractors
- American Fire Sprinkler Association
- Carolinas Associated General Contractors
- Metal Building Manufacturers Association
- National Association of Minority Contractors
- National Association of Women in Construction
- National Insulation Association
- Painting and Decorating Contractors of America

Some of the features of the *Wheels of Learning* program include:

- A proven record of success over many years of use by industry companies
- National standardization providing "portability" of learned job skills and educational credits that will be of tremendous value to trainees
- Recognition: Upon successful completion of training with an accredited sponsor, trainees receive an industry-recognized certificate and transcript from NCCER.
- Approved by the U.S. Department of Labor for use in formal apprenticeship programs
- Well illustrated, up to date, and practical. All standardized manuals are reviewed annually in a continuous improvement process.

Contents

Basic Safety

Module 00101

NATIONAL
CENTER FOR
CONSTRUCTION
EDUCATION AND
RESEARCH

BASIC SAFETY

Objectives

Upon completion of this module, the trainee will be able to:

1. Describe how to avoid job-site accidents.

2. Explain the relationship between housekeeping and safety.

3. Appreciate the importance of following all safety rules and company safety policies.

4. Explain the importance of reporting all on-the-job injuries, accidents, and near misses.

5. Explain the need for evacuation procedures and the importance of following them.

6. Explain their employer's substance abuse policy and how it relates to their safety.

7. Use proper safety practices when welding or working around welding operations.

8. Use proper safety practices when working in or near trenches and excavations.

9. Explain the term Proximity Work.

10. Follow safe practices when working near pressurized or high-temperature systems.

11. Know and follow the safety requirements for working in confined spaces.

12. Explain and practice safe lockout/tagout procedures.

13. Know the different types of barriers and barricades, and where they should be used.

14. Recognize and explain personal protective equipment uses.

15. Inspect and care for various types of personal protective equipment.

16. Follow safe procedures for lifting heavy objects.

17. Inspect and safely work with various types of ladders and scaffolds.

18. Demonstrate an understanding of the OSHA Hazard Communication Standard.

19. Explain the function of Material Safety Data Sheets.

20. Explain the process by which fires start.

21. Practice fire prevention in dealing with various flammable materials.

22. Explain the classes of fires, and the type(s) of extinguishers to use for each.

23. Explain why injuries result when electrical contact occurs.

24. Practice safe work procedures around electrical hazards.

25. Take action if present when an electrical shock occurs.

Prerequisites

None.

How to Use This Manual

During the course of completing this module, you will be taught and will practice basic safety of the construction trade. *Self-Check Review / Practice Questions* will follow the introduction of most topics. The answers to these written exercises are found in Appendix A of this manual, titled *Answers to Self-Check Review / Practice Questions*.

New Terms will be introduced in **bold** print. The definition of these terms can be found in the front of this manual, under *Trade Terms Introduced in This Module*.

Required Student Materials

1. Company safety policies
2. Site evacuation procedures
3. Locks and safety tags
4. Typical barriers and barricades
5. Typical hard hats
6. Typical safety glasses and goggles
7. Typical safety harnesses
8. Typical work gloves
9. Typical safety shoes
10. Typical ear protectors
11. Sample respiratory protection equipment
12. Assorted objects for practice lifting
13. Portable straight ladders
14. Extension ladders
15. Stepladders
16. Scaffolding
17. Sample Material Safety Data Sheets (MSDSs)
18. Representative job site fire extinguishers
19. Typical approved electrical extension cords
20. Typical approved temporary lighting equipment
21. Approved portable power tools with 3-wire cords
22. Approved, double-insulated portable power tools

Course Map Information

This course map shows all of the *Wheels of Learning* task modules in the Core Curricula. The suggested training order begins at the bottom and proceeds up. Skill levels increase as a trainee advances on the course map. The training order may be adjusted by the local Training Program Sponsor.

Course Map: Core Curricula, Basic Safety

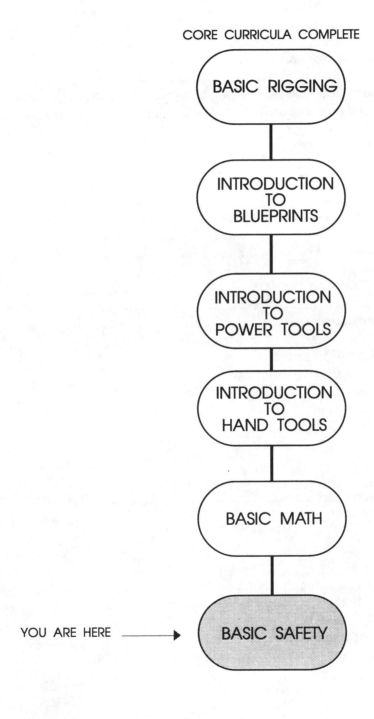

CORE CURRICULA COMPLETE

BASIC RIGGING

INTRODUCTION TO BLUEPRINTS

INTRODUCTION TO POWER TOOLS

INTRODUCTION TO HAND TOOLS

BASIC MATH

YOU ARE HERE ⟶ BASIC SAFETY

TABLE OF CONTENTS

Trade Terms Introduced In This Module

Confined space: Any area that has limited exits and has toxic or flammable materials or limited amounts of air to breathe, such as storage tanks, sewers, and tunnels.

Extension ladder: Two straight ladders assembled such that the overall length of the combination can be adjusted.

Hazard Communication Standard (HazCom): The Occupational Safety and Health Administration standard which requires all contractors to educate employees about hazardous chemicals and the methods to work safely in their presence.

Lanyard: A short section of rope used to attach a worker's safety harness to a strong anchor point located above the work area.

Lockout/Tagout: An established, formal procedure for deactivating equipment and systems and making them safe for work.

Material Safety Data Sheet (MSDS): A document accompanying any materials which contains the identity of the substance, exposure limits, its physical and chemical characteristics, the nature of the hazard it presents, precautions for safe handling and use, and specific control measures.

OSHA: The Occupational Safety and Health Administration; an agency of the U.S. Department of Labor.

OSHA Recordable: An injury, accident, or illness which must be reported under the provisions of the Occupational Safety and Health Act of 1970 by recording it on the employer's injury and illness log 200.

Personal Protective Equipment: Pieces of equipment or clothing designed to prevent or reduce the effect of injuries.

Proximity Work: Working near a hazard, but not actually contacting the source of the hazard.

Respirator: A device designed to provide clean, filtered air for breathing no matter what is in the surrounding air.

Scaffold: An elevated work platform for both personnel and materials.

Scaffolding: A manufactured or job-built structure that supports a work platform.

| **Stepladder:** | A self-supporting ladder consisting of two elements hinged at the top. |
| **Straight Ladder:** | A non-adjustable ladder of a fixed length. |

1.0.0   INTRODUCTION

The final responsibility for on-the-job safety rests with you. In this module, you will learn how to ensure your safety and the safety of the people you work with by:

1. Following safe work practices and procedures;
2. Inspecting safety equipment before use; and,
3. Using safety equipment properly.

Construction sites can be hazardous places to work. But a thorough understanding of the information provided in this module will help you avoid injury and accident.

2.0.0   ACCIDENTS: CAUSES AND RESULTS

You might hear someone on the job say, "I want my company to have a perfect safety record." However, a safety record is more than the number of days a company has worked without an accident. Safety is an attitude. Safety is a way of working on the job. The time spent learning and practicing safety procedures can literally save your life – and the lives of others.

What makes accidents happen? Accidents are caused by either poor human behavior or poor work conditions. Most accidents can be prevented – through safe work habits and through understanding what causes accidents. As the National Safety Council states, the organized safety movement has saved three million lives or, more accurately, prevented three million tragic early deaths.

Accidents cost billions of dollars each year and cause much needless human suffering. In this section of the *Basic Safety* Module, we will examine why accidents happen and how – with your help – they can be prevented.

The procedures, facts, and skills you will learn in this module will help you work safely. You will be able to spot and avoid hazardous conditions on the job site. By practicing safety procedures and keeping an attitude of safety, you will do your part to keep your workplace free from accidents and to protect yourself and your fellow workers from harm.

2.1.0 WHAT CAUSES ACCIDENTS?

You may already know some of the main reasons that accidents occur. They include the following:

1. Failure to communicate
2. Poor work habits
3. Drug or alcohol abuse
4. Lack of skill

Each of these causes will be discussed briefly below.

2.1.1 Failure To Communicate

Many accidents happen because of a lack of communication. For example, you may learn how to do things one way on a particular job, but what happens when you go on to a new job site? You need to communicate with the people at the new job site. You need to find out if they do things the way you have learned to do them. Accidents can result from communication failure because specific procedures and practices are different among individuals, companies and job sites.

If you believe that other people know something without talking with them about it, then you assume they know. Assuming that other people will do what you think they should is a very common problem. And it is the cause of many accidents.

WARNING!   **NEVER ASSUME ANYTHING!** It never hurts to ask. However, it can be a disaster if you don't ask. For example, do not assume or believe that electrical current is turned off before you touch the wires. First, ask if the current is off – that is, communicate first.

2.1.2 Poor Work Habits

Many serious accidents are caused by poor work habits. Some examples of these bad habits are simple carelessness, horseplay, or procrastination. Procrastination, or putting things off until "tomorrow," is a common cause of accidents. Putting off the repair, inspection, or cleaning of equipment and tools, for example, can cause trouble. Those who try to extend the capabilities of machines and equipment beyond their operating capacities are subjecting themselves and their fellow workers to possible injury.

Did you know that more accidents are caused by dull blades than by sharp ones? If you fail to keep your cutting tools properly sharpened, they won't cut very easily. When you have a difficult time cutting, you exert more strain, force, and pressure on the tool. When that happens, something is bound to slip. And, when something slips, you get cut.

Work habits and work attitudes are closely related. If you resist taking orders, you may also resist words of warning. If you let yourself be easily distracted, you won't be able to concentrate. If you aren't concentrating, you could make some costly mistakes.

Machines, power tools, even a pair of pliers, can cause you harm if you don't treat them with the respect that they deserve. Tools and machines don't know the difference between wood and steel and flesh and bone.

Figure 1. Horseplay Can Be Dangerous!

The activity in *Figure 1* may look like fun. It's not! In fact, it's potentially lethal. If you horse around on the job, or if you don't think what you are doing demands complete concentration, you are demonstrating a poor work attitude. This can disrupt what you're doing to such an extent that it may cause an accident.

2.1.3 Alcohol and Drug Abuse

Alcohol and drug abuse costs the construction industry millions of dollars a year in accidents, lost time, and lost productivity.

The true cost of drug and alcohol abuse is much more than just money, of course. Abuse can cost lives. As sure as drunk driving kills thousands on our highways every year, alcohol and drug abuse kills on the construction site. How would you like to be the person in *Figure 2*? Would you like to be working near him?

How many unsafe practices
can you find in this picture?

Figure 2. Drugs or Alcohol and Construction Work – A Deadly Combination

Using drugs or alcohol exposes everyone on a job site to increased chances of an injury. Many states now have laws that prevent workers from collecting insurance benefits if they are injured while under the influence of alcohol or illegal drugs.

Would you trust your life to a crane operator who was stoned? Would you bet your life on the responses of a co-worker on drugs or alcohol? The simple fact is that alcohol or drug abuse has no place in the construction industry. A person on a construction site who is under the influence of drugs or alcohol is an accident waiting to happen – a potentially fatal accident.

Studies indicate that eleven to fourteen percent of Americans abuse drugs and/or alcohol. Obviously, working in an impaired or mentally altered condition not only puts these people at risk of accident or injury, it risks their fellow workers as well. For this reason, your employer probably has a formal substance abuse policy. You should be aware of that policy and adhere to it for your own safety.

You don't have to be abusing illegal substances such as marijuana, cocaine, or heroin to present a job hazard. Many prescribed and over-the-counter drugs, being taken for legitimate reasons, can affect your ability to safely work. Amphetamines, barbituates, and antihistamines are only a few of the legitimate drugs that can affect your ability to work safely in construction, operate machinery, etc. If your physician is prescribing any medication that you feel might affect your job performance, ASK. Your safety and the safety of your fellow workers relies on everyone being alert and attentive on the job.

Many do not realize that alcohol is a drug. Even a small amount of alcohol can affect judgement, balance, and motor skills. On a construction job site, this can be deadly.

Do yourself and the people you work with a big favor. Be aware of and follow your employer's substance abuse policy. Avoid any consumption of substances that can affect your job performance. The life you save could be your own!

2.1.4 Lack of Skill

You should learn and practice new skills under careful supervision. Never perform them alone until you've been checked out by a supervisor.

Lack of skill can cause accidents quickly. Here's an example: You are told to cut a number of 2 X 8s with a circular saw, but you aren't skilled with that tool.

A basic rule of circular saw operation is to never cut without a properly functioning guard. Because you lack the skill and training, you don't know this.

You find that the guard on the saw is slowing you down. So, you jam the guard open with a small block of wood. The result could be a serious accident.

Never operate a power tool until you have been trained to use it. You can greatly reduce the chances of accidents by learning safety rules for each task that you perform.

2.2.0 HOUSEKEEPING

Trade housekeeping means more than simply keeping your work area clean and free of pipe and metal scraps or spilled water and oil. It also means orderliness and organization. Materials and supplies should be safely stored and properly labeled. Tools and equipment should be arranged to permit safe, efficient work practices and easy cleaning.

If indoors, the work site should be well-lighted and ventilated. Aisles and exits should be free of materials and obsolete equipment. Flammable liquids should be stored in safety cans. Oily rags should be placed only in approved self-closing metal containers. Keep in mind that the major goal of effective housekeeping is to prevent accidents. Good housekeeping reduces the chances for falls, slips, fires, explosions, and falling objects.

Good housekeeping rules include:

1. All scrap material and lumber with protruding nails must be kept clear of work areas.
2. All combustible scrap materials must be removed regularly.
3. Containers must be provided for both collection and separation of refuse. Those for flammable or harmful refuse must be covered.
4. All wastes must be disposed of frequently.
5. All tools and equipment should be stored when you're finished using them.

Another way of saying this is Pride of Workmanship. If you take pride in what you are doing, you simply will not allow trash to build up around you.

The old adage "A place for everything, and everything in its place" may sound corny, but it's the right idea just the same.

2.3.0 COMPANY SAFETY POLICIES

All construction companies have written policies and procedures covering safety practices on their job sites. Adherence to these policies and procedures is MANDATORY.

Be smart, like the worker in *Figure 3*. Be sure your supervisor makes you aware of the policies in effect at your job site and how you can gain access to them.

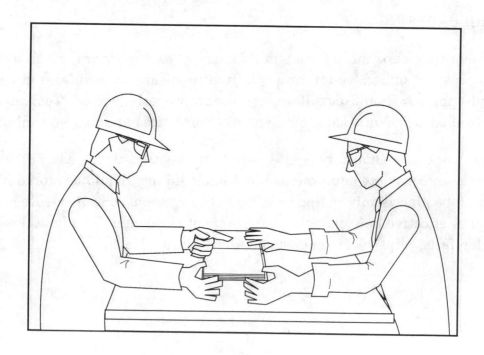

Figure 3. Learn Your Company's Policies and Procedures

2.4.0   RULES OF BEHAVIOR

Your safety is affected not only by the manner in which you do your work, but also by the way you conduct yourself on the job site. For this reason, most company policies include very strict requirements for employee behavior. Horseplay and other inappropriate behavior is strictly forbidden – with termination frequently the result.

These strict policies are for YOUR protection. The many hazards on construction sites require that an individual's behavior, whether at work, on a break, or eating lunch, is always consistent with the principles of safety.

2.5.0   REPORTING INJURIES, ACCIDENTS, AND NEAR MISSES

It is crucial that all on-the-job injuries, no matter how minor, be reported to your immediate supervisor. Many workers feel that this gets them in trouble and tend to avoid reporting minor injuries. Nothing could be further from the truth. There have been many cases of minor injuries, such as cuts and scrapes, later becoming major as a result of infection and other complications. The injured parties' Workman's Compensation payments were delayed and sometimes denied. Why? Because the accident which caused the injury had not been reported.

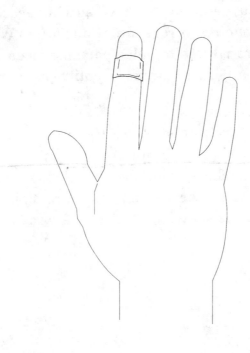

Figure 4. ALL Injuries Must Be Reported

It is by analysis of accidents that safety policies and procedures are developed, revised, and improved. Reporting an accident can and does result in procedural changes when needed. Also, good procedures reduce the likelihood of similar accidents occurring in the future.

Certain on-the-job incidents must be recorded and reported to **OSHA**, the Occupational Safety and Health Administration. These incidents include:

1. Injuries;
2. Accidents; and,
3. Near Misses.

An injury is anything that requires first aid treatment.

An accident is any occurrence that causes an injury.

A near miss is any occurrence that COULD have caused an injury but, because it was caught in time, did not.

Your company's safety policies and procedures are designed to meet the needs of OSHA's regulations on reportable incidents.

2.6.0 EVACUATION PROCEDURES

In many work environments, it is necessary to have specific evacuation procedures. These procedures are put into effect when dangerous situations arise. When this is the case, it is imperative that you know the procedures to follow for an evacuation. You must also know the signal that is used, usually a horn or siren, to notify personnel that an evacuation is required.

On hearing the evacuation signal, you must follow the procedure TO THE LETTER. That usually means following a prescribed route from your location to a designated assembly area. And you must notify the person in charge when you reach the assembly area. In the event of atmospheric release of hazardous materials, the procedure may call for visual observation of a wind sock. A typical wind sock is shown in *Figure 5*. A wind sock tells you the direction the wind is blowing.

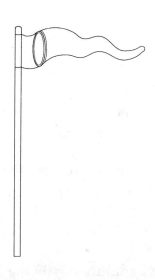

Figure 5. Wind Sock

Different evacuation routes are prescribed for different wind directions. This minimizes your exposure to the released material.

SELF-CHECK REVIEW / PRACTICE QUESTIONS 1

1. Who is affected when a construction site accident occurs?

2. List at least three causes of job site accidents.

3. Which causes more accidents, a dull blade or a sharp blade? Explain your answer.

4. Whose safety is affected by a worker using drugs or alcohol on the job?

5. Is it important to report a minor on-the-job injury? Why or why not?

PERFORMANCE / LABORATORY EXERCISES

1. During a job site tour or at your facility, point out to your instructor at least one example of good safety practices being applied.

2. During a job site tour or at your facility, point out to your instructor at least one example of good housekeeping being practiced.

3. During a job site tour or at your facility, point out to your instructor at least one example of good work habits being followed.

3.0.0 CONSTRUCTION SITE JOB HAZARDS

It would be impossible to list ALL of the potential hazards that can exist on a construction job site. In this section, you will learn about some of the more common hazards and how to deal with them. You may want to make a list of other hazards that you feel could be present on a job site and discuss them with your instructor or supervisor.

It is imperative for your safety that you know the specific hazards present where you are working and how to practice the necessary actions to prevent accidents and injuries.

3.1.0 WELDING

Even if you're not welding, you can be injured by being around a welding operation. Oxygen and acetylene used in gas welding are highly dangerous. The transport, storage, and handling of cylinders containing these substances must be done with extreme care.

The following safety guidelines should always be followed for welding safety:

1. Keep the work area clean and free from dangerous material.
2. Always handle compressed gas cylinders with extreme caution.
3. Never stare at an arc welding operation with the naked eye. Sometimes, even a reflected arc can cause burns to the eye.
4. If you are welding, use the proper protective equipment. This equipment includes the following:

    - lens shields
    - long-sleeved shirts
    - leather welders' gloves
    - high-top welders' shoes
    - cuffless trousers which cover your ankles and shoe tops
    - respiratory protection, if necessary.

5. If you are welding and other workers are in the area around your work, set up welding shields. Make sure everyone wears flash goggles.
6. A welder must be protected when his shield is down. A helper or watcher must protect him or rope off the area.
7. Welded material is hot! Mark it with a clear sign. And stay clear for a reasonable length of time after the welding has been completed.

Number 3 on the above list bears extra attention. Even a brief exposure to the ultraviolet light from arc welding can do extreme damage to your eyes. The symptoms frequently do not occur until well after the exposure. Symptoms of flash burns to the eye include the following:

- headache
- red or weeping eyes
- trouble opening the eyes
- impaired vision
- swelling of the eyes

If you feel you may have experienced a flash burn to the eyes, you should seek medical attention immediately.

WARNING!   Wearing contact lenses while welding is strictly prohibited.

3.2.0 TRENCHING AND EXCAVATING

Many construction jobs include the need for trenches and excavations. Safe work procedures must be followed by those performing this work, and by any others in the area, to prevent accident and injury. Cave-ins and objects and tools falling into an excavation are constant threats. Good job safety around trenches and excavations includes the following:

1. Never put tools, materials, or loose dirt or rocks near the edge of a trench. They can easily fall and injure the people in the trench. Also, excessive weight near the edge of a trench could cause a cave-in.
2. Always walk around a trench. Never jump over or straddle it. You could lose your footing and fall in. Or, your weight could cause a cave-in.
3. Never jump into a trench. Always use a ladder to get in and out.
4. Place barricades around all trenches, as shown in *Figure 6*.

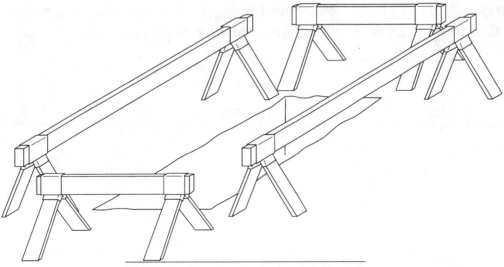

Figure 6. Barricade Around a Trench

5. Always follow OSHA regulations and your employer's procedures for shoring and securing a trench or excavation from cave-in. Never work beyond the shoring. Check the excavation daily, especially after a rain.

3.3.0 PROXIMITY WORK

Working near a hazard but not in direct contact with it is called **Proximity Work**. Proximity work requires extra procedures, caution, and awareness. The hazard may take the form of hot piping, energized electrical equipment or apparatus, running motors or machinery, etc. In any case, you must conduct yourself and perform your work such that you do not come into contact with the hazard.

You may need to erect barricades to prevent accidental contact. Lifting and rigging operations may have to be conducted in such a way as to minimize the risk of dropping material or equipment on the hazard-containing structure. In some cases, a "watcher" may be called for to monitor your work and alert you in advance to any potential contact. Energized electrical equipment is particularly hazardous. Regulations and procedures spell out the minimum safe working distance from energized electrical conductors. The distance that must be maintained varies with the voltage levels involved.

You'll learn more about working with and around energized electrical equipment in a later section of this module.

3.4.0 PRESSURIZED OR HIGH-TEMPERATURE SYSTEMS

Many construction tasks require that work be performed in close proximity to tanks, piping systems, pumps, and other components that contain pressurized and/or high temperature fluids. There are two potential hazards:

1. Simply contacting a container of high temperature fluid can result in burns. *(See Figure 7.)* Many industrial processes contain fluids whose temperatures are several hundred, if not thousand, degrees. Contact with components of these processes can cause severe burns very quickly.

Figure 7. Avoid Contacting High Temperature Components

CORE CURRICULA TRAINEE TASK MODULE 00101

2. Mechanical damage to components carrying pressurized fluids can result in leaks and dangerous fluids spraying into the area. *(See Figure 8.)*

Figure 8. Avoid Damage to Systems Containing High Pressure Fluids

Any work around pressurized or high-temperature systems should be considered proximity work and dealt with accordingly. Barricades and/or a "watcher" may be needed to ensure safety. *(See Figure 9.)*

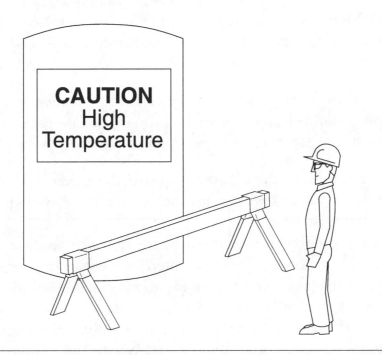

Figure 9. Work Safely Near Pressurized or High-Temperature Systems

3.5.0 CONFINED SPACES

Construction and maintenance work isn't always done in the great outdoors. Much of it takes place in **confined spaces** such as tanks and trenches. A confined space is any area where the atmosphere isn't easily ventilated. *(See Figure 10.)*

OSHA defines a confined space as:

"Confined or enclosed space" means any space having a limited means of egress, which is subject to the accumulation of toxic or flammable contaminants or has an oxygen deficient atmosphere. Confined or enclosed spaces include, but are not limited to, storage tanks, process vessels, bins, boilers, ventilation or exhaust ducts, sewers, underground utility vaults, tunnels, pipelines, and open top spaces more than 4 feet in depth such as pits, tubs, vaults, and vessels.

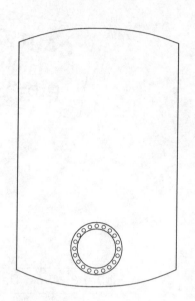

Figure 10. Typical Confined Space

Many confined spaces normally contain hazardous gases and/or fluids when the equipment is in operation. In addition, the work you are doing may introduce hazardous fumes into the space. Welding is an example of this. Special precautions are needed before entering a confined space and while working in a confined space to ensure safety.

When your work involves a confined space, you should always follow your employer's prescribed procedures. Confined space procedures may include receiving clearance from a safety representative before starting work, and any time the job conditions change.

At a minimum, working in a confined space requires the following:
1. Stay in voice or visual contact with someone outside the confined area. If you are the attendant or fire monitor, stay alert.
2. Know your company's emergency procedures for confined space work.
3. Check that air sample readings have been taken for levels of oxygen and explosive gases. Respiratory protection equipment may be required. You'll learn more about respiratory protection equipment in a later section of this module.
4. Use only approved electrical appliances, lights, extension cords, and tools.
5. Never enter a confined space in an emergency. Notify the proper personnel.
6. Use safety harness equipment.

3.6.0 MOTORIZED VEHICLES

Many motorized vehicles are found on job sites. Trucks, forklifts, backhoes, cranes, and trenchers are just a few. Working safely around this equipment requires care on the part of equipment operators. Care must also be used by helpers, riggers, or anyone else working in the vicinity.

If the equipment is used indoors, ventilation of the work area is especially important. All internal combustion engines give off carbon monoxide as part of their exhaust. Carbon monoxide is a tasteless, odorless gas. But, in high concentrations, it is LETHAL. Adequate ventilation must be ensured before operating any motorized vehicle in an indoor environment.

The operator of any vehicle is responsible for the safety of passengers and the protection of any load. The following safety guidelines should be followed when operating vehicles on a job site:

1. Always wear seat belts.
2. Obey all speed limits. Reduce speed in crowded areas.
3. Look to the rear and sound the horn before backing up. If rear vision is blocked, a signalman should direct the operator.
4. Make sure the back-up alarm is functioning.
5. **Always shut off the engine during fueling.**
6. Turn off the engine and set the brakes before leaving the vehicle.
7. All vehicles used to carry people should have a firmly secured seat for each person carried.
8. The maximum number of people allowed in the front seat of any vehicle is dependent upon the number of seatbelts.
9. Always think safety when operating any motorized vehicle.
10. Never remain on or in a truck that is being loaded by excavating equipment.
11. Windshields, rear-view mirrors, and lights should be kept clean and functional.
12. Vehicles should carry road flares, fire extinguishers, and other standard safety equipment at all times.

Cranes and other vehicles used to lift loads require additional precautions. A signalman must be used whenever the operator cannot see in the direction of the vehicle's travel or the position of the load. A clear method of communication must be established between the operator and the signalman. Neither of them can allow any distractions.

SELF-CHECK REVIEW / PRACTICE QUESTIONS 2

1. List at least four typical construction site job hazards.

2. When do the symptoms of flash burns to the eyes occur?

3. List at least three items of protective equipment which must be worn when welding.

4. Is it OK to jump over a trench as long as no one is working in it? Explain your answer.

5. What is Proximity Work?

6. Is work around a high-temperature system Proximity Work? Explain your answer.

7. What is a Confined Space?

8. List at least three precautions which should be followed when working in a Confined Space.

9. Who is responsible for protection of a load being hauled by a motorized vehicle?

10. Which type or types of motorized vehicles must have back-up alarms?

11. When must a signalman be used for operation of a motorized vehicle?

PERFORMANCE / LABORATORY EXERCISES

1. During a site tour, point out at least two welding operations to your instructor.

2. During a site tour, point out at least one example of trenching or excavation to your instructor.

3. During a site tour, point out at least one example of Proximity Work to your instructor.

4. During a site tour, point out at least one example of work being done around pressurized or high temperature systems to your instructor.

5. During a site tour, point out at least one example of a Confined Space to your instructor.

4.0.0　WORKING SAFELY WITH JOB HAZARDS

All of the job hazards you have learned about can be handled safely, but only if proper procedures are followed. As long as safety procedures are followed and a proper safety attitude is maintained, you need not fear for your well-being on the job site. It is imperative that you and your fellow workers follow approved procedures for a safe work environment to be maintained. In this section, you will learn about procedures and safety equipment commonly used on construction sites to ensure worker safety.

4.1.0　LOCKOUT/TAGOUT

A **lockout/tagout** system is followed to protect workers from potential hazards. These systems protect you from hazards such as the following:

- electricity
- flammable liquids
- chemicals
- acids
- steam
- machinery
- hydraulics
- high temperatures
- air pressure

When someone is working on or with any of the above, components or systems are shut down, drained, and/or de-energized to maintain the safety of the workers. Tags and locks are used to ensure that the status of the system or component doesn't change while work is being performed. That is, locks and tags are used to make sure that motors aren't started, valves aren't opened or closed, or any other changes are made which would compromise worker safety. Tags and locks are placed on each switch, circuit breaker, valve, or other component that is operated to isolate the equipment. In this way, workers are protected from all potential sources of energy including electrical, mechanical, hydraulic, thermal, pneumatic, and high temperature. Generally, each lockout has its own key, and the key is kept by the person who places the lock. The person who installs a lock is the only one who can remove it. Tags, with the words "Danger" or "Clearance" are also frequently used at points of isolation. *(See Figure 11.)*

Figure 11. Typical Safety Tags

A safe lockout/tagout system will be ensured if you follow these simple rules:

1. All electrical systems require lockout as well as tagging.
2. NEVER operate any device, valve, switch, or piece of equipment with a tag, or lock and tag attached to it.
3. Use only those tags which have been approved for use on your job site.
4. If a device, valve, switch, or piece of equipment is locked out, the proper tag should also be attached.
5. For any mechanical repair on motorized vehicles and equipment, the equipment must be tagged out before starting work. The starting devices should be disconnected or disabled.
6. Pipelines containing acids, explosive fluids, and high pressure steam require lockout as well as tagging.

The exact procedures followed for lockout/tagout vary between different companies and job sites. It is your responsibility to KNOW the lockout/tagout procedure in use on your job site and to follow it. This is for your safety and the safety of your fellow workers.

4.2.0 Barriers and Barricades

Any opening in a wall or floor is a safety hazard that must be recognized. There are two basic ways to protect yourself and others from these hazards. The openings may be guarded or covered. Any hole in a floor should be covered when possible. Use barricading when covering is not practical. If the bottom of a wall opening is less than three feet above the floor and would allow for a fall of four feet or more, it must be guarded. Several different protection methods may be used.

1. RAILINGS are used across wall openings or as a barrier to floor openings to prevent falls.
2. WARNING BARRICADES alert others to hazards in the area, but provide no real protection. Typical warning barricades consist of plastic tape or rope strung from wire or between posts. The tape or rope is colored-coded, according to its function:

   • Red means danger is present. No one can enter an area with a red warning barricade. This type of barricade is used when there is danger from falling tools or objects, or when a load is suspended over an area.
   • Yellow means caution. You can enter an area with a yellow barricade. However, know what hazard is present, and conduct yourself safely. Yellow barricades are frequently used around wet areas, or areas containing loose dust. Yellow with black lettering marks physical hazards such as striking against, stumbling, or falling
   • Yellow and magenta (a color like purple) means radiation warning. No entry is allowed beyond a yellow and magenta barricade. These are most commonly found when piping welds are being X-Rayed.

3. PROTECTIVE BARRICADES provide a visual warning and protection from injury. They can be wooden posts and rails, posts and chain, steel cable, or other types. Protective barricades physically prevent a person from passing beyond them.
4. BLINKING LIGHTS are frequently placed on barricades to ensure that they will be seen at night.
5. HOLE COVERS are used to cover open holes, such as in a floor. They must be strong enough to support the weight of anything that may be placed on them.

The types of barriers and barricades used vary from site to site. The procedures for when and how they are erected also vary. Be sure you learn and follow the policies in force at your job site.

WARNING!   NEVER remove a barricade unless you have been properly authorized to do so.

5.0.0 PERSONAL PROTECTIVE EQUIPMENT

Personal protective equipment includes pieces of safety-related equipment designed to protect you from injury. These items will prevent injuries only if you keep them in good condition and use them as required. Many on-the-job injuries are caused by workers not using personal protective equipment.

5.1.0 PERSONAL PROTECTIVE EQUIPMENT NEEDS

Just looking around a construction site does not reveal all of the potentially dangerous conditions. It's important to stop and consider what type of accidents COULD happen on any job you are about to do. Using common sense and knowing how to use personal protective equipment will reduce your chance of an injury.

5.2.0 PERSONAL PROTECTIVE EQUIPMENT USE AND CARE

The best protective equipment available is of no value to you unless you do four things.

1. Regularly inspect it.
2. Properly care for it.
3. Use it properly when needed.
4. Never alter or modify it in any way.

In the sections that follow, protective equipment commonly found on construction sites will be explained and proper use and care will be described.

5.2.1 Hard Hat

Figure 12 shows a typical Hard Hat. The outer shell of a hard hat is made to protect the head from a hard blow. The webbing inside the hat maintains a space between the shell and your head. The headband lets you adjust the fit so that the webbing fits your head comfortably. The webbing should be adjusted so that there is at least one inch of space between it and the shell.

At one time, hard hats were made of metal. However, because metal will conduct electricity, most hard hats are now made of reinforced plastic or fiberglass.

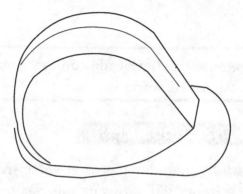

Figure 12. Typical Hard Hat

Your hard hat should be inspected every time it is used. If there are any cracks or dents in the shell, or if any webbing straps are worn or torn, get a new hard hat. Wash the webbing and headband with soapy water as often as needed to keep them clean.

5.2.2 Safety Glasses, Goggles, and Face Shields

Eye protection such as the examples in *Figure 13* should be worn wherever there is even the slightest chance of an eye injury. Areas where there are potential eye hazards from falling or flying objects are usually identified. But you should always be on the lookout for possible hazards.

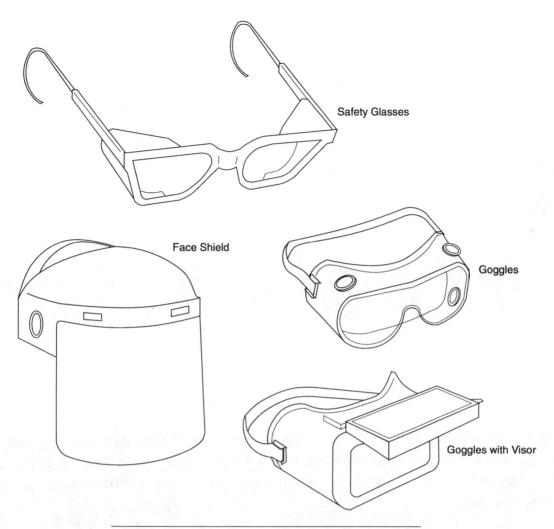

Figure 13. Typical Safety Glasses and Goggles

Regular safety glasses will protect you from objects falling or flying at you from the front. Side shields can be added to give protection from the sides. In some instances, face shields may be required. The best protection from all directions is provided by safety goggles.

Welders must use tinted goggles or welding hoods. The tinted lenses prevent the bright welding arc or flame from harming the eyes.

WARNING! Wearing contact lenses while welding is strictly prohibited. Safety glasses and goggles must be handled with care. If they become scratched, replace them – the scratches will impair your vision. Clean the lenses regularly with lens tissues or a soft cloth.

5.2.3   Safety Harness

Safety harnesses such as shown in *Figure 14* are extra-heavy duty harnesses that buckle around your upper body. They have leg, shoulder, chest, and pelvic straps.

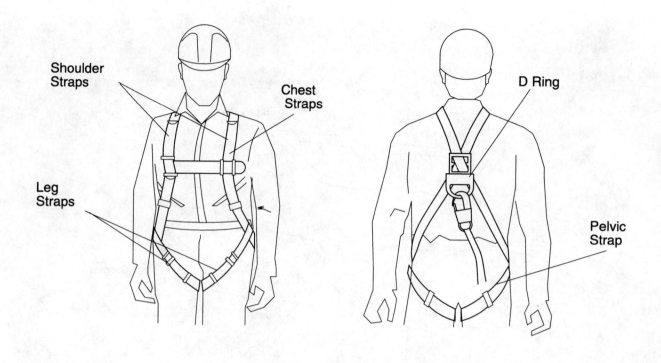

Figure 14. Typical Safety Harness

Safety harnesses have a D-Ring used to attach a short section of rope called a **lanyard** (*see Figure 15*). The lanyard is also attached to a strong anchor point located above the work area. The lanyard should be long enough to allow you to work, but short enough to limit any fall to six feet.

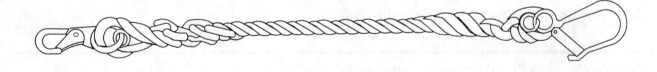

Figure 15. Lanyard

A safety harness and lanyard should be used in the following situations:

1. When you are working more than six feet above ground.
2. When you are working near a large opening in a floor.
3. When you are working near a deep hole.
4. When you are working near protruding re-bar.

The lanyard should always be attached above your head.

A safety harness and lanyard must NEVER be used for anything other than their intended purpose.

Treat a safety harness as if your life depends on it – it often does! Carefully inspect the harness before each use. Check that the buckles and D-Ring are not bent or deeply scratched. Check the harness for any cuts or rough spots. If you find any damage, turn the harness in for testing or replacement.

5.2.4 Gloves

Many construction jobs require the use of heavy-duty gloves to prevent injury to the hands (see Figure 16). Cloth, canvas, and leather are the most common materials used for construction work gloves. Never use cloth gloves around rotating or moving equipment.

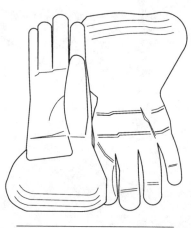

Figure 16. Work Gloves

Gloves are used to prevent cuts and scrapes when handling sharp or rough materials. Heat-resistant gloves are sometimes used for handling hot materials. Special rubber insulated gloves are used by electricians when working on or around live circuits.

When gloves become worn, torn, or soaked with oil or chemicals, they should be replaced and disposed of properly. Electricians' rubber-insulated gloves should be tested regularly to make sure they will protect the wearer from high voltage.

5.2.5   Safety Shoes

The best shoes to wear on a construction site are steel-toe, steel-sole safety shoes. The steel toe protects your toes from falling objects. The steel sole prevents nails and other sharp objects from puncturing your foot. The next best footwear material is heavy leather. Canvas shoes or sandals should NEVER be worn on a construction site.

Always replace boots or shoes when the sole tread becomes worn or when there are holes in the upper part. Oil-soaked shoes should not be worn while welding because of the risk of fire.

5.2.6   Hearing Protection

Pain is felt when most parts of the body are injured. However, the ear does not always give this warning when it is being damaged. Exposure to high noise levels over a long period of time can cause loss of hearing, even though the noise is not loud enough to cause pain.

Most construction companies follow OSHA rules in deciding when ear protection must be used. Ear protection is usually provided by using ear plugs or ear muffs specially designed for this purpose. Ear plugs fit into the ear and are made to filter out the noise in the area (see Figure 17).

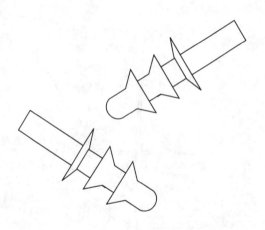

Figure 17. Ear Plugs for Hearing Protection

Ear muffs are large padded covers for the entire ear. (See Figure 18.) Depending on the noise level present, one type of hearing protection may not be adequate. To reduce noise to an adequate level both ear plugs and ear muffs may be required.

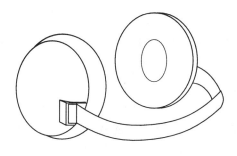

Figure 18. Ear Muffs for Hearing Protection

Ear plugs should be cleaned regularly with soap and water to prevent ear infection. The headband on ear muffs must be adjusted to provide a snug fit.

5.2.7 Respiratory Protection

Wherever there is danger of suffocation or other breathing hazards, the use of a **respirator** is required. Federal law dictates which type of respirator should be used to protect the worker from different types of hazards. There are four general types of respirators:

- Self-contained breathing apparatus (SCBA)
- Supplied air
- Full facepiece mask with chemical cannister (Gas Mask)
- Half mask or mouthpiece with mechanical filter

Self-contained breathing apparatus, or SCBA, has its own air supply carried in a compressed air tank. It may be used where there is a lack of oxygen or where there are dangerous gases or fumes in the air.

A supplied air mask uses a remote compressor or air tank. A hose supplies air to the mask. Supplied air masks may be used under the same conditions as SCBA.

Full facepiece masks with chemical cannisters are normally used for brief exposure to dangerous gases or fumes.

A half mask or mouthpiece with a mechanical filter is used only in areas where dust or other solid particles might be inhaled.

Note: Local and OSHA procedures must be followed when selecting the proper type of respirator for a particular job.

It is very important that a respirator be carefully checked for damage, and for a proper fit. A leaking facepiece can be as dangerous as no respirator at all.

Respirators used by only one person should be cleaned after each day of use, or more often if necessary. Those used by more than one person should be cleaned AND DISINFECTED (made germ-free) after each use.

Note: Usually, when a respirator is required a personal monitoring device is also required, for example, a carbon monitor.

6.0.0   LIFTING

It may surprise you, but 25% of all occupational injuries occur when handling or moving construction materials. Many of these occur when lifting heavy objects. Like most things, there is a right and wrong way to lift an object. Lifting the wrong way can land you in the hospital. The proper way to lift a heavy object is shown in *Figure 19*.

Figure 19. How To Lift Safely

CORE CURRICULA TRAINEE TASK MODULE 00101

Step 1   Move close to the object to be lifted.

Step 2   Position your feet in a forward/backward stride, with one foot at the side of the object.

Step 3   Bend your knees and lower your body, keeping your back straight and as nearly upright as possible.

Step 4   Place your hands under the object, wrap your arms around it, or grasp any provided handles. To get your hands underneath an object flat on the floor, use both hands to lift one corner of the object. Then slip one hand under it. With one hand under, you can tilt the object to get the other hand under the opposite side.

Step 5   Draw the object close to your body.

Step 6   Lift by slowly straightening your legs and keeping the object's weight as much as possible over your legs.

Step 7   Pick the object up in the direction of travel to prevent twisting your knees or back.

Following this procedure allows you to use your strongest muscles (those in your legs) rather than your weakest ones (those in your back) to lift. Practice it first with light objects. Then when you've got it down, move on to heavier ones.

Back Belts: many employers are supplying back belts as a way to reduce back injuries. However, back belts should only be used if you are trained in their proper use. A back belt is never a substitute for using the proper lifting techniques.

SELF-CHECK REVIEW / PRACTICE QUESTIONS 3

1. Who keeps the key to a lock used for lockout/tagout?

2. What types of tags may be used for lockout/tagout?

3. Can you operate a tagged valve if you get permission from the person that hung the tag? Explain your answer.

4. What's the difference between a warning barricade and a protective barricade?

5. Are all potentially dangerous conditions on a job site visually obvious? Explain your answer.

6. What should you do if you find the lenses of your safety glasses scratched?

7. When should safety harnesses be used?

8. The sole tread on your work shoes is worn, but there are no holes in the soles. Is it safe to wear them?

9. Which type or types of respirators have clean air supplied to them?

10. Safe lifting of heavy objects requires that the lifting be done primarily with which muscles?

PERFORMANCE / LABORATORY EXERCISES

1. Inspect a hard hat and determine whether or not it is safe to use.
2. Inspect safety glasses and/or goggles and determine whether or not they are safe to use.
3. Inspect safety harnesses and lanyards and determine whether or not they are safe to use.
4. Inspect gloves and determine whether or not they are safe to use.
5. Inspect safety shoes and determine whether or not they are safe to use.
6. Inspect ear protectors and determine whether or not they are safe to use.
7. Inspect respiratory protection equipment and determine whether or not it is safe to use.
8. Demonstrate the ability to safely lift a heavy object.

7.0.0   AERIAL WORK

Working in elevated locations is common in the construction industry. And, if done properly and using the proper equipment, it is done safely. Falls from heights can have devastating results. Serious injuries and even death can be the consequences. In this section, you will learn about the equipment commonly used for aerial work. You'll learn how to use it, how to inspect it, and how to maintain it.

7.1.0   LADDERS AND SCAFFOLDS

Ladders and **scaffolds** are used to perform work in an elevated location. Any time work is performed above ground level, there is an increased chance that an accident will occur. You can reduce this accident risk by carefully inspecting ladders and scaffolds before use, and by using them properly. Your knowledge of the safe operation of ladders and scaffolds will contribute to your on-the-job safety.

7.1.1   Portable Straight Ladders

Figure 20 shows a typical portable **straight ladder**.

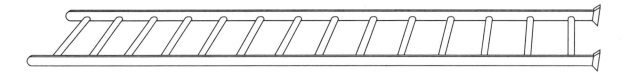

Figure 20. Portable Straight Ladder

Straight ladders consist of two rails, rungs between the rails, and safety feet on the bottom of the rails. The straight ladders used in construction are made of wood or fiberglass. Metal ladders should never be used. Metal will conduct electricity. A metal ladder could become a hazard if used around electrical equipment. Dry wood and fiberglass will not conduct electricity.

7.1.2   Inspecting Straight Ladders

Ladders should always be inspected before use. Examine the rails and rungs for cracks or other damage. Also check for loose rungs. If any damage is found, the ladder should not be used. OSHA requires regular inspections of all ladders, and an inspection immediately before each use.

Note: Wooden ladders should NEVER be painted. The paint could hide
cracks in the rungs or rails. Clear varnish, shellac, or a preservative
oil finish will protect the wood without hiding defects.

Figure 21 shows the safety feet attached to a straight ladder. Make sure the feet are securely attached and that there is no damage or excessive wear. Do not use a ladder if its safety feet are not in good working order.

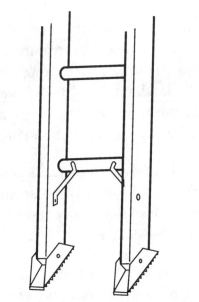

Figure 21. Ladder Safety Feet

Check the entire ladder for loose nails, screws, brackets, or other hardware. If any hardware problems are found, tighten loose parts, or have the ladder repaired before using it.

7.1.3   Using Straight Ladders

It is critically important that you place a straight ladder at the proper angle before use. An improperly angled ladder will be unstable, and you will run the risk of the ladder falling. *Figure 22* shows a properly positioned straight ladder.

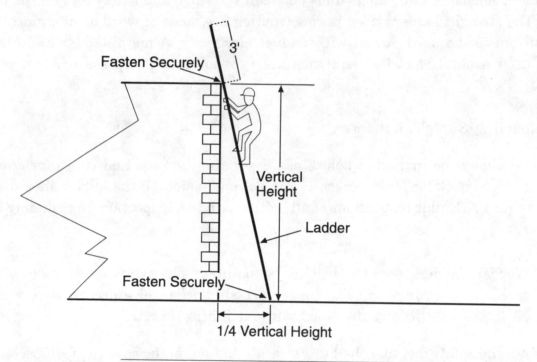

Figure 22. Proper Positioning of a Straight Ladder

The distance between the bottom of a ladder and the base of the structure it is leaning against must be one-fourth of the distance between the ground and the point where the ladder contacts the structure. For example, if the vertical height of the wall shown in *Figure 22* was 16 feet, you would position the ladder's feet 4 feet from the base of the wall. If you are going to step off a ladder onto a platform or roof, the top of the ladder should extend at least 3 feet above the point where the ladder contacts the structure.

Ladder should be used only on stable and level surfaces unless they are secured to prevent any accidental movement. Ladders should always be secured at the top and bottom. If a ladder must be placed in front of a door that opens toward the ladder, the door should be locked or blocked open. Otherwise, the door could be opened into the ladder.

Ladders are made for vertical use only. NEVER use a ladder as a work platform by placing it horizontally. Never try to move a ladder while you are on it.

CORE CURRICULA TRAINEE TASK MODULE 00101

When climbing a straight ladder, keep both hands on the rails. Always keep your body's weight in the center of the ladder between the rails. Face the ladder at all times. Never go up or down a ladder while facing away from it. If you will use a tool while on a ladder, use a hand line or tag line. Then, climb to the required height and pull the tool up. Don't carry tools in your hands while climbing a ladder.

7.1.4 Extension Ladders

An **extension ladder** is actually two straight ladders. The two are assembled using a mechanism which allows the overlap between them to be varied. This makes it possible to vary the overall length of the ladder to match the needs of a particular job. *Figure 23* shows a typical extension ladder.

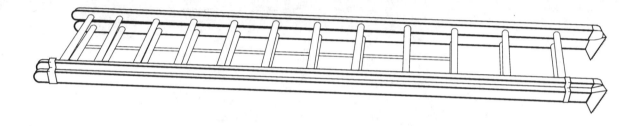

Figure 23. Typical Extension Ladder

7.1.5 Inspecting Extension Ladders

The rules for inspecting straight ladders also apply to extension ladders. In addition, you should also inspect the rope that is used to raise and lower the movable section of the ladder. If it is frayed or has worn spots, it should be replaced before the ladder is used.

The rung locks support the entire weight of the moveable section and the person climbing the ladder. They should be inspected for damage before each use. If damage is found, they should be repaired or replaced before the ladder is used.

7.1.6 Using Extension Ladders

Extension ladders are positioned and secured following the same rules as for straight ladders. When adjusting an extension ladder's length, always reposition the moveable section from the bottom, not the top. The rung locks cannot always be seen from the top. Be sure they are properly engaged after making an adjustment.

To ensure its strength, an extension ladder needs a certain amount of minimum overlap between its sections. For ladders up to 36 feet long, the overlap must be at least 3 feet. Ladders 36 to 48 feet long require at least 4 feet of overlap. Five feet is needed for those 48 to 60 feet long.

7.1.7 Stepladders

Stepladders are self-supporting ladders made up of two sections hinged at the top. A typical example is shown in *Figure 24*.

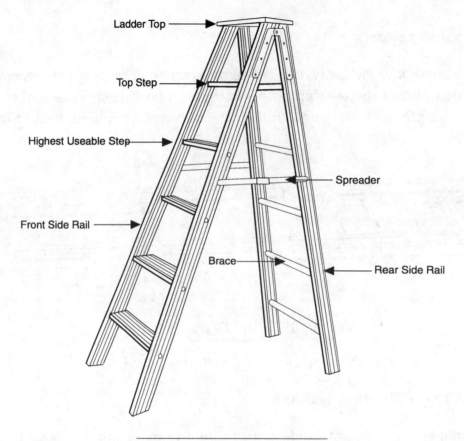

Figure 24. Typical Stepladder

The section used for climbing consists of rails and rungs similar to those found on straight ladders. The other section consists of rails and braces. The braces should NEVER be used for climbing. Spreaders are hinged arms between the sections which position the ladder for stability and prevent it from folding while in use.

7.1.8 Inspecting Stepladders

Stepladders should be inspected following the procedure used for straight and extension ladders. Pay particular attention to the hinges and spreaders to be sure they are in a good state of repair. Also examine the rungs for cleanliness. Since a stepladder's rungs are usually flat, oil, grease, or dirt can accumulate on them. Be sure the rungs are clean. Any accumulation of dirt or other material could interfere with safe footing.

7.1.9 Using Stepladders

When placing a stepladder in position, be sure that all four feet are on a hard, even surface. If they're not, the ladder can rock from side to side or corner to corner when you climb it. With the ladder in position, be sure the spreaders are locked in the fully open position. When climbing a stepladder, never stand on the top step or the top of the ladder. The top step is so near the top that putting your weight this high on the ladder will make it unstable. The top of the ladder is made to support the hinges, NOT to be used as a step. Although the rear braces may look like rungs, they are not designed to support your weight. NEVER climb the back of a stepladder. For certain jobs, however, specially designed two-person ladders are used with steps on both sides.

7.2.0 SCAFFOLDS

Scaffolds provide safe, secure elevated work platforms for personnel and materials. Scaffolds are designed and built in compliance with high safety standards. But normal wear and tear or accidental overstress can weaken a scaffold and make it unsafe to use. Inspection of all parts of a scaffold before each use is therefore very important.

There are three basic types of scaffolds used in the construction industry and the rules for safe use apply to all of them. The three types of scaffolds are manufactured, suspended, and rolling.

7.2.1 Manufactured Scaffolds

Manufactured scaffolds are made of painted steel, stainless steel, or aluminum. This makes them stronger and more fire-resistant than wooden scaffolds. They are supplied in ready-made units which resemble the sections of a fence. The individual units are assembled on site. *Figure 25* shows a typical manufactured scaffold assembled for use.

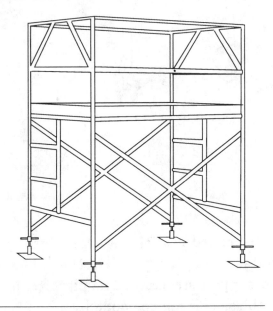

Figure 25. Typical Manufactured Scaffold

7.2.2 Rolling Scaffolds

A rolling scaffold is essentially a manufactured scaffold that has wheels on its legs so that it can be easily moved. *Figure 26* shows a typical example. The scaffold wheels are fitted with brakes to prevent movement while work is in progress.

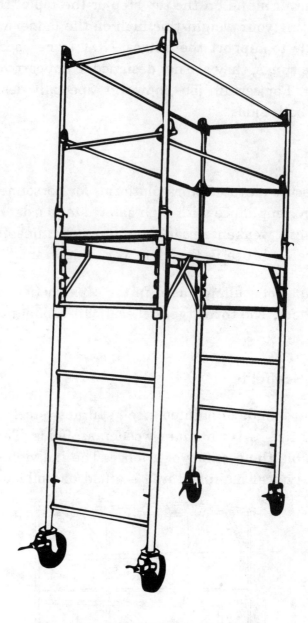

Figure 26. Typical Rolling Scaffold

7.2.3 Suspended Scaffolds

As shown in *Figure 27*, a suspended scaffold is a platform supported by ropes or cables attached to the top of some support structure. Or the cables may be attached to beams extending out of the side of a support structure. The platform is raised or lowered by pulling on the suspension ropes or cables. This may be done by a hand crank or by an electric motor.

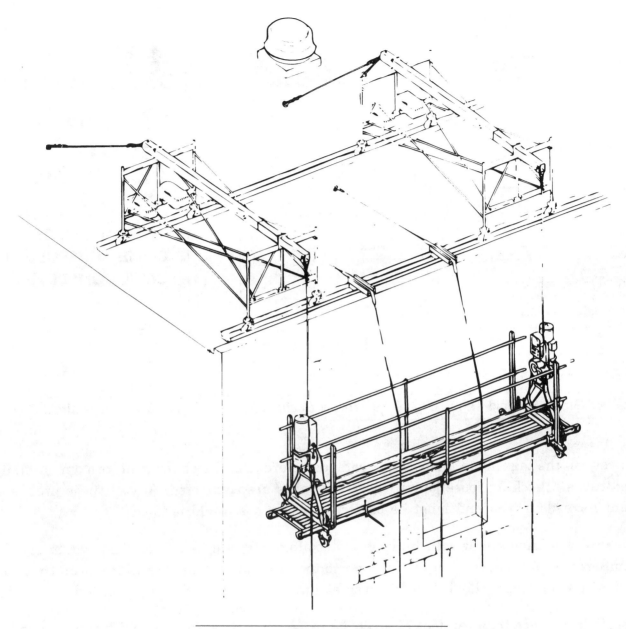

Figure 27. Typical Suspended Scaffold

7.2.4 Inspecting Scaffolds

Any scaffold assembled for use should be tagged. Three colors of tags are used: green, yellow, or red.

A green tag identifies a scaffold which is safe for use. It meets all OSHA standards.

A yellow tag means the scaffold does NOT meet all applicable standards. An example is a scaffold where a railing cannot be installed because of equipment interference. A yellow-tagged scaffold may be used. However, a safety harness and lanyard is mandatory. Other precautions may also apply.

A red tag means a scaffold is being erected or taken down. You should never use a red-tagged scaffold.

Figures 28, 29, and 30 show typical scaffold tags.

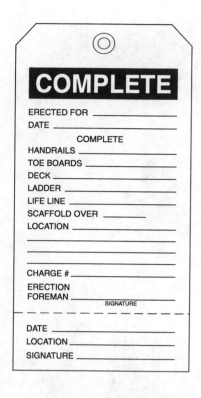

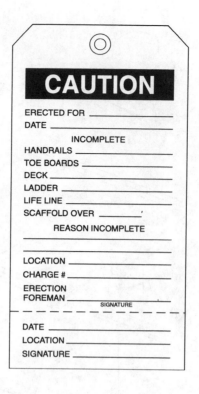

Figure 28. Green Scaffold Tag   Figure 29. Yellow Scaffold Tag   Figure 30. Red Scaffold Tag

Don't rely on the tags alone. Inspect all scaffolds before use. Check for bent, broken, or badly rusted tubes. Check for loose joints where the tubes are connected. Any of these problems present hazards to you and must be corrected before a scaffold is used.

Make sure you know the weight limit of any scaffold you will be using. This weight should be compared with the total weight of the personnel, tools, equipment, and material that you expect to put on the scaffold. Scaffold weight limits must NEVER be exceeded.

If a scaffold is more than ten feet high, check to see that it is equipped with top rails, mid-rails, and toe boards. All connections must be pinned. Cross bracing must be used. The working area must be completely planked. Cross bracing is NOT a handrail.

If it is possible for people to pass under a scaffold, the space between the toe board and top rail must be screened. This prevents tools and work materials from falling off the work platform.

When a manufactured or rolling scaffold is taller than four times its narrow base dimension, it should be tied to the structure or guyed (wired) to the ground. For example, if a scaffold with a base of four by six feet is taller than 16 feet, it should be tied or guyed.

When inspecting a rolling scaffold, check the condition of the wheels and brakes. Be certain that the brakes are working properly and are capable of preventing the scaffold from moving while work is in progress. And be sure all brakes are locked before using the scaffold.

A suspended scaffold should be inspected while it is on the ground. Then, raise the scaffold about one foot while two persons are standing in the center of the work platform. It should be stable and balanced, and should not drift downward. If ANY drifting occurs, its cause must be determined and corrected before the scaffold is used.

WARNING!   Unless specially made, suspended scaffolds are not meant to support more than two persons.

7.2.5   Using Scaffolds

Be sure that there is firm footing under each leg of a scaffold before placing any weight on it. If necessary, two-inch thick lumber may be placed under the scaffold's legs or wheels when working on loose or soft soil.

When it is necessary to move a rolling scaffold, the following sequence must ALWAYS be used:

Step 1   Dismount.

Step 2   Unlock the wheels.

Step 3   Move the scaffold.

Step 4   Re-lock the wheels.

Step 5   Remount.

WARNING!   NEVER unlock a rolling scaffold's wheels while someone is on the scaffold.

Wear a safety harness whenever working on a suspended scaffold. The lanyard should be attached to an independent life line, not to the scaffold.

Never use a scaffold suspended by Manila rope under any of the following conditions:

• When working with acids
• When welding or flamecutting
• When abrasive blasting.

All of these can damage or weaken the rope. Wire rope is required for suspended scaffolds used for doing these jobs.

SELF-CHECK REVIEW / PRACTICE EXERCISE 4

1. What should you look for when inspecting a straight ladder?

2. You are leaning a straight ladder against the top of a wall. How far should the base of the ladder be from the base of the wall?

3. Under what conditions should a straight ladder or extension ladder be secured? Where should it be secured?

4. Is it safe for two persons to work from a stepladder as long as one is in front and the other in back?

5. When and why must scaffolds have toe boards?

6. What is indicated by a scaffold's weight limit?

PERFORMANCE / LABORATORY EXERCISES

1. Inspect a portable straight ladder and determine if it is safe to use.

2. Inspect an extension ladder and determine if it is safe to use.

3. Inspect a stepladder and determine if it is safe to use.

4. Inspect a manufactured scaffold and determine if it is safe to use.

5. Inspect a rolling scaffold and determine if it is safe to use.

6. Inspect a suspended scaffold and determine if it is safe to use.

8.0.0 HAZARD COMMUNICATION STANDARD (HAZCOM)

The U.S. Department of Labor's Occupational Safety and Health Administration (OSHA) has a standard which affects every worker in the construction industry. The rule is called the **Hazard Communication Standard (HazCom)**. You may have heard this document referred to as the "Right To Know" requirement. It addresses the worker's right to know the specifics about any hazardous materials they may come into contact with on the job. The standard requires all contractors to educate their employees about the hazardous chemicals they may be exposed to on the job site. Employees must be taught the methods to work safely in the presence of these materials.

Many people believe there are few hazardous chemicals on construction job sites. Nothing could be farther from the truth. In the standard, the term "hazardous chemical" applies to paint, concrete, and even wood dust.

8.1.0 MATERIAL SAFETY DATA SHEETS (MSDSs)

A **Material Safety Data Sheet (MSDS)** should accompany every shipment of a hazardous substance and be available to you on the job site. *Figure 31* shows a portion of a typical MSDS.

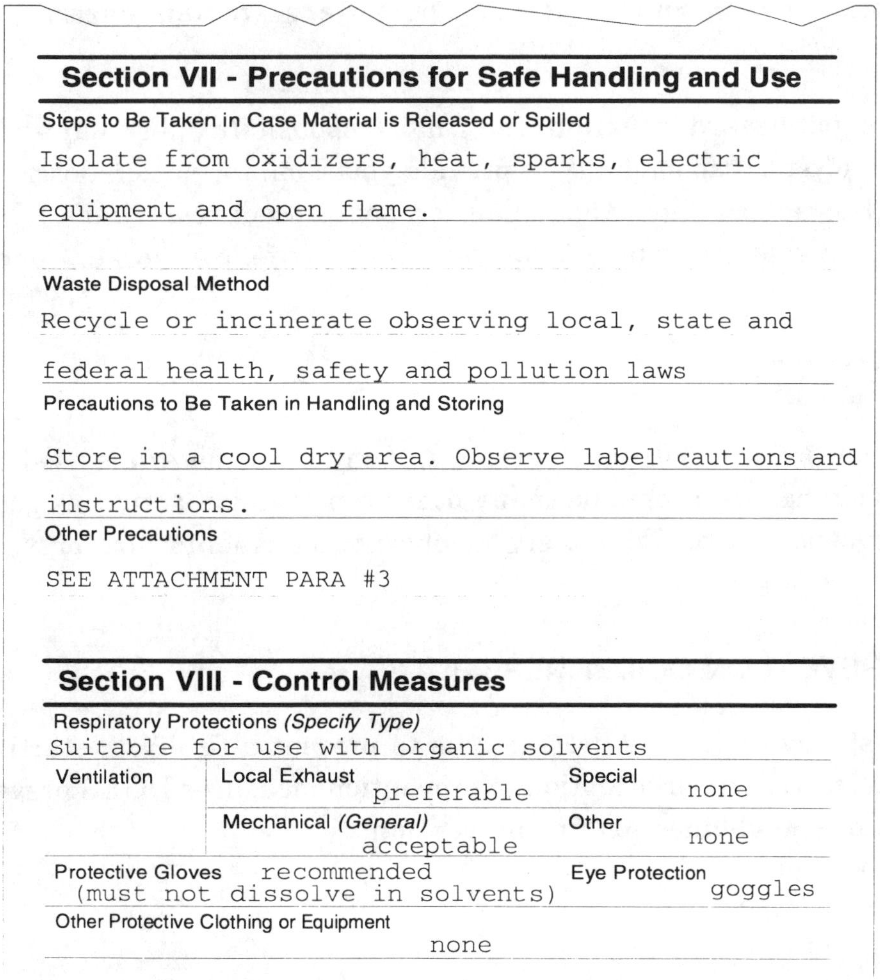

Section VII - Precautions for Safe Handling and Use

Steps to Be Taken in Case Material is Released or Spilled

Isolate from oxidizers, heat, sparks, electric equipment and open flame.

Waste Disposal Method

Recycle or incinerate observing local, state and federal health, safety and pollution laws

Precautions to Be Taken in Handling and Storing

Store in a cool dry area. Observe label cautions and instructions.

Other Precautions

SEE ATTACHMENT PARA #3

Section VIII - Control Measures

Respiratory Protections *(Specify Type)*
Suitable for use with organic solvents

| Ventilation | Local Exhaust preferable | Special none |
| | Mechanical *(General)* acceptable | Other none |

| Protective Gloves recommended (must not dissolve in solvents) | Eye Protection goggles |

Other Protective Clothing or Equipment
none

Figure 31. Typical MSDS

The information on an MSDS includes the following:

- The identity of the substance
- Exposure limits
- Physical and chemical characteristics
- The nature of the hazard it presents
- Precautions for safe handling and use
- Reactivity
- Specific control measures

8.2.0 YOUR RESPONSIBILITIES UNDER HAZCOM

Your responsibilities under HazCom may be summarized as follows:

1. Know the location of MSDSs on your job site.
2. Spot and report hazards on the job site.
3. Know the physical and health hazards of any hazardous materials on your job site.
4. Know and practice the actions necessary to protect yourself from these hazards.
5. Know the actions necessary in an emergency.
6. Know the location and content of your employer's written hazard communication program.

As with any other job hazard situation, the final responsibility for your safety rests with you. It is your employer's responsibility to provide you the information described above. But you must know this information and follow proper procedures. That's what makes the difference between a safe environment for you and your fellow workers and an injury or accident.

9.0.0 FIRE SAFETY

Fire is a constant hazard on construction job sites. Many of the materials used in construction are flammable. Welding, grinding, and many other construction activities generate heat or sparks that could cause a fire. Fire safety involves two elements: fire prevention and fire fighting.

9.1.0 FIRE PREVENTION GUIDELINES

Obviously, the best way to provide fire safety is to prevent a fire from starting in the first place. And, in most cases, proper attention to prevention measures DOES prevent fires. Here are some basic safety guidelines for fire prevention.

1. Always work in a well-ventilated area, especially when using flammable materials such as shellac, lacquer, paint stripper, construction adhesives, etc.
2. Never smoke or light matches when working with flammable materials.
3. Keep oily rags in approved self-closing metal containers.
4. Store combustible materials only in approved containers.
5. Always know the location of fire extinguishers, what kind to use for different kinds of fires, and how to use them.
6. Make sure all extinguishers are fully charged. Never remove the tag from an extinguisher – it indicates the last date the extinguisher was serviced and inspected.

9.2.0   HOW FIRES START

For a fire to start, three things are needed in the same place at the same time: fuel, heat, and oxygen. If any of these three is missing, a fire cannot be started. And, if a fire has started, removing any one or more of these from the fire will extinguish it.

FUEL is anything that will combine with oxygen when heat is present. When pure oxygen is present, such as near a leaking oxygen hose or fitting, substances that would not normally be considered fuels (including some metals) will burn.

HEAT, for these purposes, is anything that will raise a fuel's temperature to the flash point. The flash point is the temperature at which a fuel gives off enough gases to burn. The flash point temperatures of many substances are quite low – room temperature or less. When the burning gases raise the temperature of a fuel to its ignition temperature, the fuel itself will burn, and keep burning, even though the original source of heat is removed.

The requirements for a fire to start are frequently shown as a fire triangle. *(See Figure 32.)* If any one element of the triangle is missing, a fire can not start. If a fire has started, removing any element from the triangle will put it out.

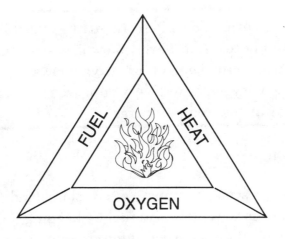

Figure 32. The Fire Triangle

9.3.0 FIRE PREVENTION

The best way to prevent a fire is to make sure that the three elements needed for fire are never present in the same place at the same time. This can be done for the different types of fuel by recognizing and removing at least one of the elements.

9.3.1 Flammable And Combustible Liquids

Flammable liquids are any liquids whose flash point is below 100 degrees Fahrenheit. Combustible liquids have a flash point at or above 100 degrees. Fire can be prevented by:

1. REMOVING THE FUEL – It is not the liquid that burns. What burns are the gases (vapors) given off as the liquid evaporates. Keeping the liquid in an approved, sealed container prevents evaporation. If there is no evaporation, there is no fuel to burn.
2. REMOVING THE HEAT – If the liquid is stored or used away from a heat source, it will not be able to ignite. There is not sufficient heat to start combustion.
3. REMOVE THE OXYGEN – The vapor from these liquids will not burn if oxygen is not present. Keeping safety containers tightly sealed prevents oxygen coming into contact with the fuel.

9.3.2 Flammable Gases

Some of the flammable gases used on construction sites are acetylene, hydrogen, ethane, and propane (LPG). To save space, these gases are compressed so that a large amount is stored in a small cylinder or bottle. As long as the gas is kept in the cylinder, both fuel and oxygen are unavailable for a fire. Storing the cylinders away from sources of heat removes heat as well.

Oxygen too is classed as a flammable gas. If it is allowed to escape and mix with another flammable gas, an explosive mixture will result.

WARNING!   NEVER use grease or oil on the fittings used for oxygen bottles and hoses. And never allow greasy or oily rags to remain near any part of an oxygen system. Oil and pressurized oxygen form a very dangerous mixture which can ignite at low temperatures.

9.3.3 Ordinary Combustibles

The term Ordinary Combustibles means paper, wood, cloth, and similar fuels. The easiest way to prevent fire in this case is to remove the fuel by keeping a neat and clean work area. If there are no scraps of paper, cloth, or wood lying around, heat and oxygen will have no fuel to start a fire. So establish and maintain good housekeeping habits. Use approved storage cabinets and containers for all waste and other Ordinary Combustibles.

CORE CURRICULA TRAINEE TASK MODULE 00101

9.4.0 FIRE FIGHTING

As a construction worker, you are not expected to be an expert fire fighter. However, you may have to deal with a fire to protect your life and your job. It's important that you know the location of fire fighting equipment on your job site. It is equally important to know which equipment to use on different types of fires.

The National Fire Protection Association (NFPA) has guidelines for how many fire extinguishers are needed on construction jobs. The guidelines also tell which type of extinguishers are needed. Most companies tell new employees where fire extinguishers are placed. If you have not been told, be sure to ask.

While you are learning the location of the fire fighting equipment, also learn the procedure for reporting fires. The telephone number of the nearest fire department should be clearly posted in your work area. If a company fire brigade has been formed, learn how to contact them in case of a fire.

9.4.1 Classes Of Fires

There are four classes of fuels that can be involved in fires. You've already learned about the first three of them. Each class of fuel requires a different method of fire fighting and a different type of extinguisher.

1. CLASS A fires are those where Ordinary Combustibles such as wood or paper are burning. A Class A fire is fought by cooling the fuel. Water is the fire fighting substance in a Class A fire extinguisher. Using a Class A extinguisher on any other type of fire can be very dangerous.
2. CLASS B fires are fires involving grease, liquids, and gases. Class B extinguishers put out these fires by cooling and smothering them. Carbon Dioxide (CO_2) or other Class B fire fighting materials work primarily by removing oxygen from the fire.
3. CLASS C fires are any fires near or involving energized electrical equipment. Class C extinguishers are designed to protect the fire fighter from electrical shock while fighting the fire. Class C extinguishers fight fire by smothering them.
4. CLASS D fires are unique. This category is reserved for a group of metals that will, in fact, burn. Class D extinguishers contain a powder which either forms a crust around the burning metal or gives off gases which prevent oxygen from reaching the fire.

Some metals will remain burning even though they have been coated with powder from a Class D extinguisher. The best way to fight these fires is to keep using the extinguisher so the fire will not spread to other fuels.

Labels clearly identify the class of fire on which each extinguisher can be used. *(See Figure 33.)*

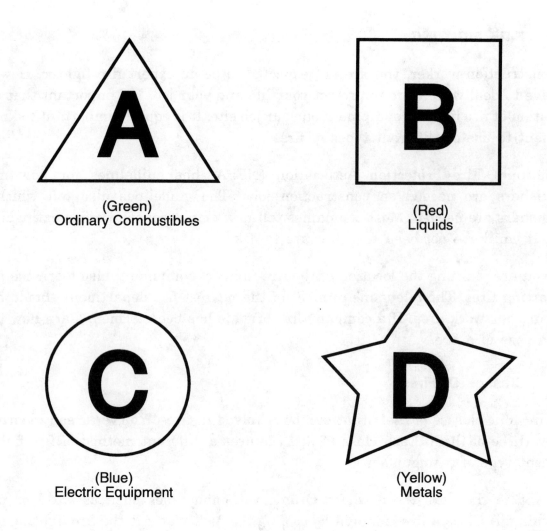

(Green)
Ordinary Combustibles

(Red)
Liquids

(Blue)
Electric Equipment

(Yellow)
Metals

Figure 33. Typical Fire Extinguisher Labels

When checking the extinguishers in your work area, you will probably notice some that are rated for more than one class of fire. An extinguisher with the codes A, B, and C may be used to fight any of the three classes of fire. But remember, if the extinguisher has only one code letter, DO NOT use it on any other class of fire – even in an emergency. You would place yourself in harm's way and, in many cases, could make the fire WORSE!

SELF-CHECK REVIEW / PRACTICE 5

1. Where is your employer's written hazard communication program on your job site?

2. What is a Material Safety Data Sheet?

3. Who is responsible for your safety under HazCom?

4. List at least four fire prevention guidelines.

5. What three things must be present in the same place at the same time for a fire to occur?

6. What is a Class A fire?

7. Is water appropriate for fighting a Class B fire? Explain your answer.

PERFORMANCE / LABORATORY EXERCISES

Safely and properly operate at least one of the fire extinguishers commonly found on construction job sites.

10.0.0   ELECTRICAL SAFETY

Many construction workers feel that electrical safety is only a concern to electricians. In fact, many of the jobs you will do, no matter what your trade, require that you use or work around electrical equipment. Extension cords, power tools, portable lights, and many other day-to-day items use electricity. And, if you don't use this equipment properly and safely, the results could be fatal – to you or someone you work with.

Electricity can be described as a potential which results in the flow of electrons through a conductor. This flow of electrons is called electrical current. Some substances, such as silver, copper, steel, and aluminium are excellent conductors. This means that electrical current flows easily through them. The human body is also a conductor. The body's skin, when dry, is not as good a conductor as the internal parts of the body. However, when moistened with perspiration, body skin is a good conductor also.

Electrical current flows along the path of least resistance back to its source. The source return point is called the neutral, or grounded conductor, of a circuit. If the human body comes in contact with an electrically energized conductor and is also in contact with the ground, it becomes the path of least resistance. When the human body acts to conduct current and the amount of current is sufficiently high, the person is said to have been electrocuted. *Figure 34* shows the effects of different amounts of current on the human body.

CURRENT FLOW	EFFECT
0.5 to 2 milliamperes	slight sensation
2 to 10 milliamperes	muscular contraction
5 to 25 milliamperes	painful shock
50 to 100 milliamperes	heart convulsions
100 or more milliamperes	breathing stops

Figure 34. Effects of Electrical Current on the Human Body

For comparison, a 100 watt light bulb uses 833 milliamperes of current.

The prime cause of death from electrical shock is the heart going into a state of convulsion called fibrillation. The normal operation of the heart has very low level electrical signals passing down and through the heart. The signals cause the heart to contract and pump blood. When an abnormal signal reaches the heart, the low level "heart beat" signals are overcome. The heart begins "twitching" in an irregular manner. This twitching is fibrillation. Fibrillation lasts for only a short time. Without the normal heart beat rhythm, the individual dies.

An example may be in order. A craftsman is operating a portable power drill while standing on damp ground. The power cord inside the drill has become frayed from hard use and it comes into contact with the drill frame. The resistance of the craftsman's body is such that 100 milliamperes of current pass through his body to ground. The table above tells us that the likely result is that his breathing stops – most likely a fatal injury.

Certainly, not all electrical accidents result in death. There are different types of electrical injuries. Any of the following can happen.

• Electric shock
• Falls caused by electrical shock
• Burns
• Explosions
• Fire

All of these are possible if proper procedures aren't followed.

Electrical burns can be extensive and deep. However, since they are sterile, they tend to heal quickly and well. More serious electrical burns may require amputation of the affected limb.

Electricity can pass for short distances through air. When it does, the arc and flash that result generate a great deal of heat. This heat can cause burns, fires, and even explosions.

10.1.0 BASIC ELECTRICAL SAFETY GUIDELINES

OSHA and your company's policies and procedures are very specific about keeping the workplace safe from electrical hazards. There are many things you can do to reduce the chance of a fatal or disabling electrical accident. Here are the basic job site electrical safety guidelines:

1. Extension cords must be the 3-wire type and protected from damage. They must never be fastened with staples, hung from nails, or suspended from wires. Never use damaged cords.
2. Panels, switches, outlets, and plugs should be grounded.

3. Never use bare electrical wire.
4. Never use metal ladders near ANY source of electricity.
5. Never wear a metal hard hat.
6. Always inspect electrical power tools before each use.
7. Never operate any electrical device that has a danger tag or lockout device attached to it.
8. Cords for portable power tools must be the 3-wire type and they must be properly connected *(see Figure 35)*.

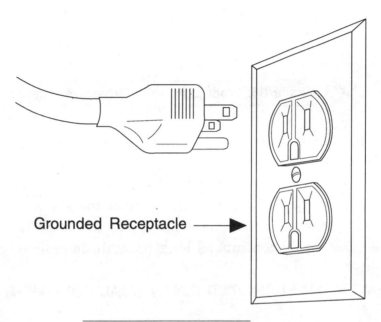

Grounded Receptacle ———▶

Figure 35. 3-Wire System

The 3-wire system is one of the most common safety grounding systems used to protect you from accidental electrical shock. The third wire is connected to ground. Should the insulation in a tool fail, the current will pass to ground through the third wire – and not through your body.

9. Worn or frayed cables must never be used.
10. Double-insulated tools designed to have a two-wire cord must be approved by Underwriters Laboratories (UL).
11. All light bulbs must have protective guards to prevent accidental contact. *(See Figure 36.)*

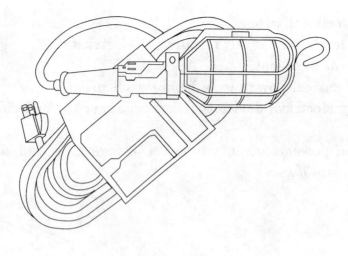

Figure 36. Work Light With Protective Guard

12. Temporary lights must not be hung by their power cords, unless they are specifically designed for this use.

13. Receptacles for attachment plugs must be of the approved, concealed type. Where different voltages or types of current are used in the same area, the receptacles must be of such design that the attachment plugs are not interchangeable.

14. Flexible cords must be in continuous lengths with no splices.

10.2.0 WORKING NEAR ENERGIZED ELECTRICAL EQUIPMENT

No matter what your trade, your job may include working near exposed electrical equipment or conductors. This is one example of Proximity Work. Many times, electrical distribution panels, switch enclosures, and other equipment must be left open during construction. This leaves the wires and components in them exposed. Some or all of the wires and components may be energized. Working near exposed electrical equipment can be safe. But only if the proper safe working distance is maintained.

Minimum safe working distances from exposed conductors are specified in regulations and company policies. They vary with the amount of voltage in the conductor. The greater the voltage, the greater the safe working distance. The required safe working distance varies from a few inches to several feet, depending on the voltage.

It is your responsibility to know the safe working distance which applies to each situation. You must conduct yourself so that you never get any part of your body or any tool you may be using closer to exposed conductors than that distance. Information on safe working distances can be obtained from your instructor, your supervisor, company safety policies, and regulatory documents.

10.3.0   IF SOMEONE IS SHOCKED

If you are present when someone receives an electrical shock, immediate action can be life-saving. Do the following:

1. Immediately disconnect the circuit.
2. If the circuit cannot be readily disconnected, use a dry board, stick, rope, coat, blanket or any other NONCONDUCTING material to separate the victim from the circuit.

WARNING!   Do not touch the victim or the electrical source unless you are safely insulated. You could become another victim.

---

3. Cut the circuit with insulated hand tools if the power cannot be otherwise disconnected.
4. Once the victim is separated from the circuit, apply first aid as required and call for an ambulance.

SELF-CHECK REVIEW / PRACTICE QUESTIONS 6

1. What is electrical shock?

2. Which trades need to be concerned about electrical safety?

3. What is fibrillation?

4. Why are 3-wire cords used on portable electrical power tools?

5. Who determines the minimum safe working distance from exposed electrical conductors?

APPENDIX A

ANSWERS TO SELF-CHECK REVIEW / PRACTICE QUESTIONS

SELF-CHECK REVIEW / PRACTICE 1

1. Everyone is affected: the injured person, his fellow workers, the employer, the insurance company, etc.
2. Failure to communicate, poor work habits, drug or alcohol abuse, lack of skill.
3. A dull blade. More force must be applied when using a dull blade, increasing the likelihood of the tool slipping.
4. The worker himself, and all his fellow workers.
5. Yes. Failure to report an injury could result in delays or denial of Workman's Compensation.

SELF-CHECK REVIEW / PRACTICE 2

1. Welding, trenching and excavating, Proximity Work, pressurized or high temperature systems, motorized vehicles.
2. Although, in extreme cases, symptoms may occur immediately, they frequently do not occur until well after the exposure.
3. Lens shields, long-sleeved shirt, welders' gloves, high-top shoes, cuffless trousers.
4. No. You could lose your footing and fall in. Or, your weight could cause a cave-in.
5. Working near a hazard, but not actually contacting the source of the hazard.
6. Yes. Since making contact with the system's components could cause burns, this would be Proximity Work.
7. Any work area where access is controlled for safety reasons.
8. Stay in voice or visual communication with someone outside, keep rescue equipment ready, check that air sample readings have been taken, use only approved electrical equipment.
9. The vehicle operator.
10. All.
11. Whenever the operator's view of the load, or in the direction of travel, is obstructed.

SELF-CHECK REVIEW / PRACTICE 3

1. The person who places the lock.
2. Only those approved for use on the job site.
3. No. The lockout/tagout procedure forbids operating any component while it is locked or tagged.
4. A warning barricade alerts you to a hazard. A protective barricade physically prevents you from passing beyond the barricade.

5. No. Although many hazardous situations are obvious, and signs or barricades may be used to identify others, not all are visually obvious. For example, noise levels or dangerous fumes cannot be seen.

6. Replace them. The scratched lenses will impair vision, and wearing them could be a safety hazard.

7. A safety harness should be used whenever you are working more than six feet above ground, near a large opening in a floor, near a deep hole, or near protruding re-bar.

8. No. Wearing them could present a hazard of slipping and falling.

9. Self-contained breathing apparatus (SCBA), and hose masks.

10. The leg muscles.

SELF-CHECK REVIEW / PRACTICE 4

1. A straight ladder should be inspected for: cracked or damaged rails or rungs; loose rungs; loose or damaged safety feet; and, loose nails, screws, brackets, or other hardware.

2. The distance between the base of the ladder and the wall should be equal to one fourth of the vertical height from the ground to the top of the wall.

3. Straight or extension ladders should always be secured top and bottom.

4. No. Two persons should never work from a stepladder at the same time. And, the back of a stepladder should never be used for climbing.

5. When it is possible for people to pass under the scaffold. Toe boards prevent tools and materials from falling off the work platform.

6. The maximum weight the scaffold can support including personnel, tools and materials.

SELF-CHECK REVIEW / PRACTICE 5

1. Your instructor will provide the answer to this question.

2. A document accompanying any hazardous materials that contains the identity of the substance, exposure limits, its physical and chemical characteristics, the nature of the hazard it presents, precautions for safe handling and use, and specific control measures.

3. Although your employer is responsible for providing information and training, the ultimate responsibility for your safety rests with you.

4. Always work in a well-ventilated area; never smoke or light matches when working with flammable materials; keep oily rags in approved metal containers; store combustible materials only in approved containers; know the location of fire extinguishers, their types, and which type to use for each type of fire; make sure all extinguishers are fully charged.

5. Fuel, heat, and oxygen.

6. One involving Ordinary Combustibles such as wood, paper, or cloth.

7. No. Class B fires involve grease, oil, or gases. Using water could spread the fire.

SELF-CHECK REVIEW / PRACTICE 6

1. The flow of electrical current through the human body.
2. All trades. No matter what the trade, they are exposed to electrical hazards.
3. A "twitching" of the heart caused by electrical shock.
4. To provide a current flow path to ground if the tool's insulation fails, and thereby prevent shock to the operator of the tool.
5. They are specified in regulations and company policies.

The NCCER makes every effort to keep these manuals up-to-date and free of technical errors. We appreciate your help in this process. If you have an idea for improving this manual, or if you find an error, a typographical mistake, or an inaccuracy in the *Wheels of Learning*, please write us, using this form or a photocopy. Be sure to include the exact module number, page number, a description of the problem, and the correction, if possible. We'll do our best to correct it in later editions. Thank you for your assistance.

Write: *Wheels of Learning*
National Center for Construction Education and Research
P.O. Box 141104
Gainesville, FL 32614-1104
Fax: 352-334-0932

WHEELS OF LEARNING USER UPDATE

Please let us know if you have found an inaccuracy, error, or other problem in a *Wheels of Learning* manual. Use this form or write us a letter. Please be sure to tell us the exact module name and module number, the page number, and the problem. Thanks for your help.

Craft   Module Name

_____

Module Number   Page Number(s)

_____

Description of Problem

_____

_____

_____

(Optional) Correction of Problem

_____

_____

_____

(Optional) Your Name and Address

_____

_____

NATIONAL
CENTER FOR
CONSTRUCTION
EDUCATION AND
RESEARCH

BASIC MATH

Objectives

Upon completion of this module, the trainee will be able to:

Work With Whole Numbers

1. Add, subtract, multiply, and divide whole numbers, with and without a calculator.

Work With Fractions And Measurement

2. Use a standard and metric ruler to measure.
3. Add, subtract, multiply, and divide fractions.
4. Add, subtract, multiply, and divide decimals, with and without a calculator.

Work With Decimals And Percents

5. Convert decimals to percents and percents to decimals.
6. Convert fractions to decimals and decimals to fractions.

Work With The Metric System

7. Explain what the Metric System is and its importance in the construction trade.
8. Recognize and use metric units of length, weight, volume, and temperature.

Prerequisites

Successful completion of the following Task Module(s) is required before beginning study of this Task Module: Task Module 00101, *Basic Safety*.

How to Use This Manual

During the course of completing this module, you will be taught and will practice using math skills required for the construction trade. *Self-Check Review / Practice Questions* will follow the introduction of most operations. The answers to these written exercises are found in Appendix A of this manual, titled *Answers to Self-Check Review / Practice Questions*.

New terms will be introduced in **bold** print. The definition of these terms can be found in the front of this manual, under *Trade Terms Introduced in This Module*.

Required Student Materials

1. Student Manual (this book)
2. Standard ruler (with 1/16" markings)
3. Metric ruler (with centimeters (cm) and millimeters (mm))
4. Machinist's rule
5. Calculator
6. Pencil and paper

Course Map Information

This course map shows all of the *Wheels of Learning* task modules in the Core Curricula. The suggested training order begins at the bottom and proceeds up. Skill levels increase as a trainee advances on the course map. The training order may be adjusted by the local Training Program Sponsor.

Course Map: Core Curricula, Basic Math

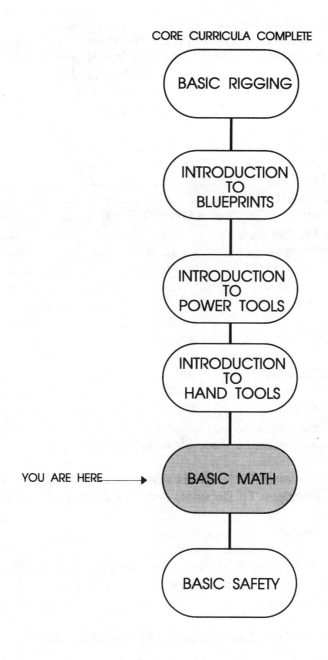

TABLE OF CONTENTS

Trade Terms Introduced in This Module:

Convert:	To change from one unit or expression to another.
Decimal:	Part of a number, represented by digits to the right of a decimal point.
Denominator:	The bottom number of a common fraction, indicates how many equal parts the whole is divided into.
Difference:	The result of a subtraction problem.
Digit:	One of the symbols 0, 1, 2, 3, 4, 5, 6, 7, 8, 9
Dividend:	In a division problem, the number being divided.
Divisor:	In a division problem, the number being divided by.
Equivalent Fractions:	Fractions having different numerators and denominators but equal value, such as 1/2 and 2/4.
Ft.	Abbreviation for "foot" or "feet."
Fraction:	A number represented by a numerator and denominator, such as 1/2.
Improper Fraction:	A fraction where the numerator is larger than the denominator.
In.	Abbreviation for "inch."
Invert:	To reverse the order or position of numbers; in common fractions to turn upside down, such as 2/3 to 3/2.
Metric Ruler:	Instrument that measures Metric measurements (millimeters, centimeters, meters).
Mixed Numbers:	Combination of whole numbers with fractions or decimals.
Negative Numbers:	Numbers less than zero.
Numerator:	The top number of a fraction, representing the number being divided.
Place Value:	The exact quantity of a digit determined by its place relative to the decimal point.
Positive Numbers:	Numbers greater than zero.
Product:	The answer in a multiplication problem.
Proper Fraction:	A fraction where the numerator is smaller than the denominator.
Quotient:	The answer in a division problem.

Remainder:	Left over amount in a division problem.
Standard English Ruler:	Instrument that measures English measurements (inches, feet, yards).
Sum:	The answer in an addition problem.
Whole Numbers:	Complete units without fractions or decimals.
'	Symbol for "feet."
"	Symbol for "inch."

1.0.0 INTRODUCTION TO WHOLE NUMBERS

Remember back in school when the teacher said, "You may need to use this mathematical operation some day"? Today is the day! In the construction trade, workers use math all the time. When you measure a length of material, fill a container with a specified amount of liquid, or calculate the dimensions of a room, you will be using math operations.

This session addresses operations on **whole numbers**. Whole numbers are complete units *without* **fractions** or **decimals**.

The following are whole numbers:

    1   5   67   335   2, 654

The following are *NOT* whole numbers:

    1/2   17/11   7-4/13   0.45   4.25

In this section of the Math module, we will be working only with whole numbers. Later we will work with the other kinds of numbers.

1.1.0 PARTS OF A WHOLE NUMBER

Let's look at the parts of a whole number.

5, 316, 247

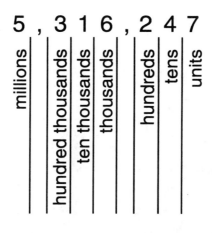

Figure 1. Place Values

The above whole number *(Figure 1)* is read as "five million, three hundred sixteen thousand, two hundred forty-seven." There are seven digits in it. Each digit represents a **place value**, depending on its place, or location, in the number.

Numbers larger than zero are called **positive (+) numbers** (such as 1, 2, 3). Numbers less than zero are called **negative (-) numbers** (such as -1, -2, -3). Zero is neither positive nor negative. Numbers without a sign in front of them are positive.

SELF-CHECK REVIEW / PRACTICE QUESTIONS 1

1.   Write this number on the line below:

      Unit's place: 4

      Ten's place: 6

      Hundred's place: 9

      Thousand's place: 3

      _3964_   (Write the number on this line.)

Look at this number:

   25, 718

What's in the:

2.   Ten's place? ____1____

3.   Thousand's? ____5____

4.   Unit's?   _____8_____

5.   Hundreds?   _____7_____

6.   Ten thousands?   _____2_____

Write the numbers for these words:

7.   Eighty-five: _____85_____

8.   One hundred twenty-two: _____122_____

9.   Two thousand four hundred ninety-seven: _____2,497_____

10.   Eighteen thousand forty-six: _____18,046_____

1.1.1   Addition Of Whole Numbers

To add means to combine two or more numbers together into a **sum** (sum is the total of an addition problem). The sign for addition is the plus sign (+). Addition problems can be written vertically (up and down) or horizontally (sideways). For example:

$$
\begin{array}{r} 6 \\ + 3 \\ \hline 9 \end{array} \qquad 6 + 3 = 9 \qquad\qquad \begin{array}{r} 5 \\ + 2 \\ \hline 7 \end{array} \qquad 5 + 2 = 7
$$

1.1.2   Carrying In Addition

Pretty easy at this point. But what happens when you want to add together larger numbers, such as:

$$
\begin{array}{r} {}^{1}58 \\ + 34 \\ \hline 92 \end{array}
$$

HOW TO CARRY IN AN ADDITION PROBLEM *(Figure 2)*

Using 58 + 34,

	THINK		WRITE	

5 tens	+ 8 units		$\overset{1}{5}8$	*1 ten carried to*
+ 3 tens	+ 4 units		+ 34	*ten's column*
(8 tens + 0)	+ (1 ten + 2 units)		92	*2 units remaining*
				in unit's column

Figure 2. Carrying in Addition

SELF-CHECK REVIEW / PRACTICE QUESTIONS 2

Time to flex your "carrying" muscles. Try these:

1.
```
    32
+   75
   107
```

2.
```
    73
+   45
   118
```

3.
```
    83
+   53
   136
```

4.
```
    452
+    74
    526
```

5.
```
     62
+   745
    807
```

6.
```
    323
+   758
  1,081
```

7.
```
  1,254
+   357
  1,611
```

8.
```
    943
+   293
  1,236
```

9.
```
  4,593
+ 8,947
 13,540
```

10.
```
  32,394
+  4,497
```

1.1.3 Problem Solving Steps

You will be using addition at work to solve construction-related problems. For example:

> *Problem*:
>
> The following lengths are cut from a bar of hot rolled steel: 19",
> 39", and 8". How many total inches of bar have been cut off?

When approached with such a problem, it may help to break it down into steps:

19"
39"
8"
66" TOTAL OR 5'6"

HOW TO SOLVE PROBLEMS

Step 1. Read the entire problem, creating a picture in your mind (or on paper) of what the problem is asking.

[For the sample problem above, picture cutting 19", 39", and 8" pieces from a bar of steel. Then picture laying them end to end to find the total amount cut.]

Step 2. Decide what it is the problem is asking.

[This problem wants to know the total number of inches of steel that have been cut off.]

Step 3. Decide what information in the problem is needed to solve it. What unit will the answer be in (for example, <u>inches</u>, feet, dollars, etc.)?

[For this problem, you need the add together the lengths of steel that have been cut off. The answer will be in inches.]

Step 4. Decide which math operation or combination of operations are needed for the solution.

[Since you are looking for the total amount of inches, you would add.]

Step 5. Estimate a reasonable answer.

[Make a guess as to the approximate number of inches. Looking at the numbers, a good guess might be 70".]

Step 6. Solve the problem.

$$
\begin{array}{r}
19'' \\
39'' \\
+\ 8'' \\
\hline
66''
\end{array}
$$

Step 7. Check the answer against your estimate. Does it make sense?

[Our estimate was 70". The actual answer is 66". The answer makes sense!]

Step 8. Check the answer against the problem. Does it answer the question? Does it make sense?

[66 total inches of bar were cut off. Thinking again of the pieces cut (19", 39", and 8"), this answer makes sense.]

SELF-CHECK REVIEW / PRACTICE QUESTIONS 3

Use all the skills you have learned so far to solve these construction-related problems.

1. A project has 8 workers on one job, 35 on another, and 18 on a third. How many people are working on these three jobs? _8 + 35 + 18 = 61 MEN_

2. A bricklayer lays 649 bricks the first day, 632 the second day, and 478 the third day. How many bricks are laid in the three days? _649 + 632 + 478 = 1759_

3. Four walls of a bathroom require 31, 46, 49, and 16 tiles. How many tiles are required for the four walls? _31 + 46 + 49 + 16 = 142_

4. In eight different boxes, there are a number of 3/4", #8 flat head, bright wood screws. The numbers of screws are 142, 57, 35, 79, 32, 79, 53, and 95. What is the total number of screws? _572_

5.   A contractor uses 36,000 bricks and orders 1,500 more to complete the job. How many bricks does the job require? _37,500_

1.1.4   Subtraction Of Whole Numbers

Subtraction means finding the **difference** between two numbers, or taking away one number from another. The subtraction sign (-) is also called the minus sign. The result (answer) of the subtraction process is called the difference.

For example:

> *Problem*:
>
> You have a total of 9 sockets to install today. You have installed 5 so far. How many more do you have to install today?
>
> 9 sockets to install today
>
> – 5 you've installed so far
>
> 4 sockets left to install

In some subtraction problems, you may have to subtract a larger digit from a smaller digit, such as:

$$\begin{array}{r} 76 \\ -\ 48 \\ \hline 28 \end{array}$$

Here, you have to subtract 8 from 6 (the digits in the units place). How can this be done? By **borrowing** from the ten's place.

HOW TO DO SUBTRACTION WITH BORROWING

PROBLEM	THINK	WRITE
76	7 tens + 6 units	76
− 48	− 4 tens + 8 units	− 48
28		

Step 1.   Notice that there are not enough units to subtract from (you cannot subtract 8 from 6).

Step 2.   Borrow 1 ten from the ten's column.

$$(6 \text{ tens} + 1 \text{ ten}) + (6 \text{ units}) \qquad 7\cancel{6}^{1}$$
$$\underline{-\quad 4 \text{ tens} \qquad\qquad + 8 \text{ units}} \quad \underline{- \ 48}$$

1 ten carries
to unit's column

Step 3.   Change the borrowed 10 to ones and add these to the units column (10 + 6 = 16). 6 tens are now left in the ten's column.

6 tens + 16 units	76
− 4 tens 8 units	−48

Step 4.   You now have enough to subtract from in each column.

6 tens + 16 units	76
− 4 tens + 8 units	− 48
2 tens 8 units	28

SELF-CHECK REVIEW / PRACTICE QUESTIONS 4

Now try these.

1. 8̸7̸
    − 38
    ___
    49

2. 2̸6̸
    − 17
    ___
    09

3. 9̸2̸
    − 34
    ___
    58

4. 24̸6̸
    − 18
    ___
    228

5. 82̸6̸
    − 717
    ___
    109

6. 46̸2̸
    − 284
    ___
    178

7. 84̸6̸
    − 218
    ___
    628

8. 3,42̸6̸
    − 2,717
    ___
    0709

9. 5,86̸2̸
    − 3,784
    ___
    2,078

CAREFUL!   Some of the following problems are *mixed* problems, using both addition and subtraction.

10.   An electrician removes from stock 250 feet of cable on Monday, 500 on Wednesday, and 750 on Thursday. On Friday, 415 feet of cable are returned. How many feet of cable were not returned? *1,085*

11.   A mason orders 64 bags of concrete but only uses 48 bags. How many bags were not used? *16*

12.   A contractor has 42 workers on one job, 26 on another, and 12 on a third. On one day, 28 workers went home with the flu. How many workers were left on the three jobs?

52 WORKERS.

1.1.5 Multiplication Of Whole Numbers

You have 8 pieces of wood. Each piece of wood requires 4 nails. How many nails will you need *(Figure 3)?*

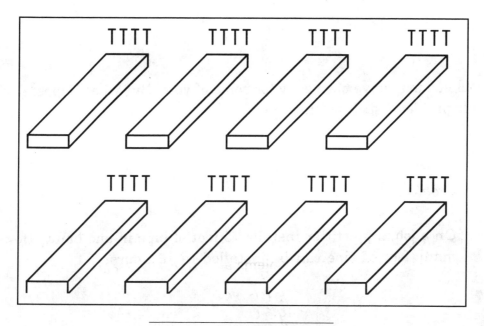

Figure 3. How Many Nails?

You *could* lay out 4 nails in a bundle 8 times and count them all. OR, you could MULTIPLY.

> 8 pieces of wood
>
> × 4 nails per piece of wood
>
> 32 nails needed

Multiplication is the quick way to add the same number together many times. In the above example, you used multiplication to figure 4 (nails) multiplied by 8 (pieces of wood), or 4 times 8.

The symbol for multiplication is the ×. The answer to a multiplication problem is called the **product**. What is the product in the problem about the nails above? The product is 32.

1. Adding 4+4+4+4+4 can be simplified to _4_ × _5_. The answer is: _20_.

2. The answer in a multiplication problem is called the _Product_.

3.   9
   × 8
   72

4.   32
   × 4
   128

5. 23
  × 2
  46

6.   12
   × 4
   48

7.   11
   × 6
   66

8. 21
  × 4
  84

The above problems may have been easy for some of you. But what happens when you have a more difficult problem, such as this one:

Problem:

On a job, a pipefitter installs 75 feet of pipe in one hour. How many feet of pipe can be installed in 16 hours?

75'
× 16 HRS
450
75
1200'

HOW TO MULTIPLY WITH LARGER NUMBERS

Multiply 75 by 16.

Step 1.   Write the numbers in 2 lines. Place units under units; tens under tens.

    75
 × 16

Step 2.   Start with the digit in the units place of the bottom number (the 6 in the 16).

    75
 × **16**

Step 3. Multiply every number in the top number (the 75) with the number in the unit's place of the bottom number (the 6). Start with the unit's place of the top number (the 5).

$$
\begin{array}{r}
3 \\
75 \\
\times\ 16 \\
\hline
450
\end{array}
$$

Six times 5 is 30, or 3 tens and 0 units. Place the 0 in the unit's place; carry the 3 tens to the ten's place.

Step 4. Now multiply every number in the top number (the 75) by the number in the ten's place (the 1) of the bottom number. Start with the number in the unit's place of the top number (the 5).

$$
\begin{array}{r}
75 \\
\times\ 16 \\
\hline
450 \\
75
\end{array}
$$

One ten times 5 units = 5 tens. Place a 5 in the ten's place.

One ten times 7 tens = 7 hundreds. Place a 7 in the hundred's place.

Step 5. Now add the numbers in each column, beginning with the unit's place.

Begin this number in the ten's place.

$$
\begin{array}{r}
75 \\
\times\ 16 \\
\hline
450 \\
+\ \ 75 \\
\hline
1,200
\end{array}
$$

The product of this problem is 1,200. This means that the answer to the problem [On a certain job, a pipefitter installs 75 feet of pipe in one hour. How many feet of pipe can be installed in 16 hours?] is that the pipefitter can install 1,200 feet of pipe in 16 hours.

Now use what you learned to solve these problems.

1.
$$\begin{array}{r} 2 \\ \times\ 5 \\ \hline 10 \end{array}$$

2.
$$\begin{array}{r} 3 \\ \times\ 5 \\ \hline 15 \end{array}$$

3.
$$\begin{array}{r} 3 \\ \times\ 3 \\ \hline 9 \end{array}$$

4.
$$\begin{array}{r} \overset{2}{4}52 \\ \times\ 4 \\ \hline 1,808 \end{array}$$

5.
$$\begin{array}{r} \overset{1}{6}2 \\ \times\ 5 \\ \hline 310 \end{array}$$

6.
$$\begin{array}{r} \overset{1\ 2}{3}23 \\ \times\ 8 \\ \hline 2,584 \end{array}$$

7.
$$\begin{array}{r} \overset{1\ 3\,2}{1,2}54 \\ \times\ 57 \\ \hline 8778 \\ 6270 \\ \hline 71,478 \end{array}$$

8.
$$\begin{array}{r} \overset{1\,2}{9}43 \\ \times\ 93 \\ \hline 2829 \\ 8487 \\ \hline 87,699 \end{array}$$

9.
$$\begin{array}{r} 4,593 \\ \times\ 47 \\ \hline 32151 \\ 18372 \\ \hline 215,871 \end{array}$$

10. A concrete-block mason lays 250 blocks a day. How many can he lay in 3 days?
    __750 BLOCKS__

11. A building uses the following size lamps: eight 15 watt, twelve 50-watt, six 25-watt, four 75-watt, and eight 100-watt. How many watts are consumed when all the lights in the building are turned on? ~~_____~~ WATTS. __1,970__

A job in your workshop requires 3 hours of work on electrical, 7 hours of work on masonry, 12 hours work on pipefitting, and 15 hours on welding. How many hours are required on each process if 94 similar jobs are ordered?

12. Electrical: __282__

13. Masonry: __658__

14. Pipefitting: __1,128__

15. Welding: __1,410__

1.1.6 Division Of Whole Numbers

Division is the opposite of multiplication. Instead of adding a number several times $(5+5+5 = 5 \times 3 = 15)$, when dividing, you subtract a quantity several times. Read this:

Problem:

You have a piece of wood that is 60" long. You need to cut pieces from it that are each 10" long. How many 10" pieces will you have *(Figure 4)*?

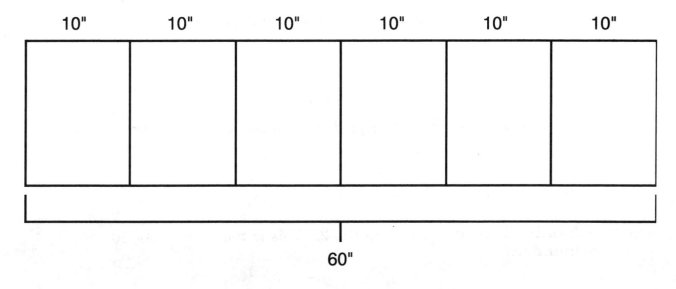

Figure 4. How Many 10" Pieces?

To solve this problem, subtract 10" from the 60" piece of wood as many times as you can.

How many times can you subtract 10" from a 60" piece of wood? Six times.

The answer in a division problem is called the **quotient**. In the problem above, the quotient is 6. The number you are dividing is called the **dividend**, which is 60 in the problem. The number you are dividing *by* is called the **divisor**, which is 10 in the problem.

This problem can be written one of two ways. Either:

$$60 \div 10 = 6 \qquad \text{or} \qquad 10)\overline{60}^{\,6}$$

What about a problem that doesn't come out even, such as:

Problem:

You have 10' of cable. You are asked to cut it up into 4' pieces. How many 4' pieces will you have? How much of the 10' cable will be left over?

10/4
02 2

WILL HAVE 2 - 4' PIECES
AND 2' LEFT OVER.

Step 1. Four goes into 10 two times. Place a 2 in the quotient.

$$\begin{array}{r} 2 \\ 4{\overline{\smash{)}}\,10} \end{array}$$

Step 2. Multiply 2 times 4. Place the product, 8, under the 10.

$$\begin{array}{r} 2 \\ 4{\overline{\smash{)}}\,10} \\ -8 \end{array}$$

Step 3. Subtract 8 from 10. The answer is 2. This is the **remainder**.

$$\begin{array}{r} 2\ r2 \\ 4{\overline{\smash{)}}\,10} \\ \underline{-8} \\ 2 \end{array}$$

You will have two 4' pieces of cable, with 2' of cable left over. This left over part is called the **remainder**.

SELF-CHECK REVIEW / PRACTICE QUESTIONS 7

Give the Quotient and remainder (if there is one) for each.

1. $15 \div 3 = $ _____5_____

2. $36 \div 4 = $ _____9_____

3. $54 \div 5 = $ _____10.8_____

4. $17 \div 6 = $ _____2.83_____

5. $39 \div 7 = $ _____5.57_____

6. $68 \div 8 = $ _____8.5_____

7.
$$\begin{array}{r} 3.25 \\ 8{\overline{\smash{)}}\,26} \end{array}$$

8.
$$\begin{array}{r} 5.83 \\ 6{\overline{\smash{)}}\,35} \end{array}$$

9.
$$\begin{array}{r} 3.6 \\ 5{\overline{\smash{)}}\,18} \end{array}$$

10.
$$\begin{array}{r} 8.75 \\ 4{\overline{\smash{)}}\,35} \end{array}$$

Now let's take a look at solving a bit more complicated division problem:

Problem:

A group of 24 people were a part of a lottery pool that just won $2,638 (after taxes). They decided to split it up evenly and whatever was left over (the remainder) would go to charity. How much did each get? How much went to charity?

$109 DID EACH GET & $22.00 CHARITY.

You could always grab a calculator. But if one wasn't handy and you had to solve it by hand - could you? You would have to use **long division**.

HOW TO DO LONG DIVISION

The problem is $2,638.00 divided by 24 people.

Step 1.   Estimate the answer. Twenty-four is close to 25. 2,638 is close to 2500. 25 goes into 2500 one hundred times. Estimate: approximately $100 for each person.

Step 2.   Now divide, using the following steps:

$$
\begin{array}{r}
1 \\
24)\overline{\,2638\,} \\
-24 \\
\hline
2
\end{array}
$$

1. 24 goes into 26 one time with a remainder of 2.

$$
\begin{array}{r}
10 \\
24)\overline{\,2638\,} \\
-24 \\
\hline
23 \\
-0 \\
\hline
\end{array}
$$

2. Bring down the next number, 3. Can 24 go into 23? It cannot, so put a 0 in the quotient.

$$
\begin{array}{r}
10 \\
24)\overline{\,2638\,} \\
-24 \\
\hline
23 \\
-0 \\
\hline
238
\end{array}
$$

3. Subtract 0 from 23. The answer is 23. Bring down the next number from the dividend, the 8.

$$\begin{array}{r} 109 \\ 24\overline{)2638} \\ -24 \\ \hline 23 \\ -0 \\ \hline 238 \end{array}$$

4. How many times can 24 go into 238? To figure this, think: 24 is close to 25. 238 is close to 250. How many times can 25 go into 250? About 9 or 10. Let's try 9.

$$\begin{array}{r} 109 \quad r22 \\ 24\overline{)2638} \\ -24 \\ \hline 23 \\ -0 \\ \hline 238 \\ -216 \\ \hline 22 \end{array}$$

5. Nine times 24 is 216. Subtract 216 from 238. The remainder is 22.

So how much money did each of the 24 people get? $109 each. How much went to charity? $22. (Note how the answer, $109, is close to your estimate of $100.)

SELF-CHECK REVIEW / PRACTICE QUESTIONS 8

Find the quotients and any remainders for the following:

1. $12\overline{)263}$ 21 R11

2. $16\overline{)4218}$ 263 R10

3. $15\overline{)4532}$ 302 R2

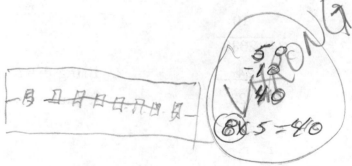

4. How many electrical outlets can be installed along a 50' wall if they are spaced 5' apart, and the first and last outlets are placed at 5' from the adjacent walls?
   9 oulets

5. A plumbing job requires 100' of plastic pipe available in 20' sections. How many sections will you need?   5 sections

6. How many rolls of insulation are in 300' of insulation, if each roll is 50' in length? _6 ROLLS_

7. A job called for 21 hours of work. It had to be done quickly, so the supervisor put 3 people on it. How many hours should it take each person to complete their share?

   7 HRS.

8. In a 144' run of cable, staples are placed 6' apart. How many staples are used if one staple is placed at the beginning and one is placed at the end of the run?

   _24_

9. A machine shop owner has $4,694 to buy tools. Each tool cost about $18. How many tools can he buy? _206_ How much money will be left? _$1.00_

10. A mason and a helper finish 2,540 square feet of concrete in 5 days. How many square feet of concrete do they finish each day? _508_

1.1.7   Using The Calculator To Add, Subtract, Multiply, And Divide Whole Numbers

It is important to be able to perform calculations (addition, subtraction, etc.) in your head even if you have access to a calculator. It allows you to estimate the answer before using the calculator. Why is this important? So that you can doublecheck the answer against your estimation. If you press a wrong number on the calculator, you might be out hundreds of dollars or off several inches.

The calculator is a marvelous tool for saving time. Let's look at the most often used operations of the calculator: how to add, subtract, multiply, and divide whole numbers.

1.1.8 Parts Of The Calculator

Below *(Figure 5)* is a drawing of a commonly used calculator.

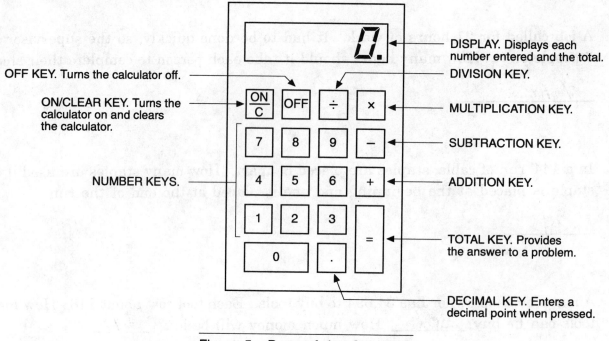

Figure 5. Parts of the Calculator

1.1.9 Using The Calculator To ADD Whole Numbers

Adding numbers is easy with a calculator. Just follow the steps below.

HOW TO ADD WHOLE NUMBERS USING A CALCULATOR

Let's use 5 + 4 to practice.

Step 1. Turn the calculator ON. A zero (0) appears in the display.

Step 2. Press 5. A 5 appears in the display.

Step 3. Press the + key. The 5 is still displayed.

Step 4. Press the 4 key. A 4 is displayed.

Step 5. Press the = key. The sum, 9, appears in the display.

Step 6. Press the ON/C key to clear the calculator.

SELF-CHECK REVIEW / PRACTICE QUESTIONS 9

Practice adding these numbers on your calculator:

1.   12 (press +)
       24 (press +)
     + 33 (press =)
     ——————
       69

2.   67
       46
     + 96
     ——————
      209

3.   83
       35
     + 50
     ——————
      168

4.   34
      938
       24
     + 63
     ——————
     1,059

5.   67
      774
      983
     + 532
     ——————
     2,356

6.   654
      543
      334
     + 552
     ——————
     2,083

7.   235
      957
     + 76
     ——————
     1,268

8.   367
      476
     + 926
     ——————
     1,769

9.   813
      135
     + 508
     ——————
     1,456

10.   327
      588
     + 65
     ——————
      980

1.1.10 Using The Calculator To SUBTRACT Whole Numbers

Subtracting with a calculator is as easy as adding with one. Here are the steps:

HOW TO SUBTRACT WHOLE NUMBERS USING A CALCULATOR

Let's use the problem 25 − 5 to practice.

20

Step 1. Turn the calculator ON. A zero (0) appears in the display.

Step 2. Press the 2 key and then the 5 key. A 25 appears in the display.

Step 3. Press the – key. The 25 is still displayed.

Step 4. Press the 5 key. A 5 is displayed.

Step 5. Press the = key. The difference, 20, appears in the display.

Step 6. Press the ON/C key to clear the calculator.

SELF-CHECK REVIEW / PRACTICE QUESTIONS 10

Use your calculator to figure these. Some may contain **mixed operations** (addition and subtraction).

1.
$$\begin{array}{r} 12 \\ -\ 5 \\ \hline 07 \end{array}$$

2.
$$\begin{array}{r} 13 \\ -\ 4 \\ \hline 08 \end{array}$$

3.
$$\begin{array}{r} 97 \\ -\ 63 \\ \hline 34 \end{array}$$

4.
$$\begin{array}{r} 452 \\ -\ 414 \\ \hline 038 \end{array}$$

5.
$$\begin{array}{r} 62 \\ -\ 25 \\ \hline 37 \end{array}$$

6.
$$\begin{array}{r} 323 \\ -\ 158 \\ \hline 165 \end{array}$$

7.
$$\begin{array}{r} 1,254 \\ -\ 557 \\ \hline 697 \end{array}$$

8.
$$\begin{array}{r} 943 \\ -\ 93 \\ \hline 850 \end{array}$$

9.
$$\begin{array}{r} 4,593 \\ -\ 4,247 \\ \hline 0346 \end{array}$$

10. At 9:00 am, the thermometer showed 88 degrees. At 1:00 pm, it read 129 degrees. How much did the temperature rise between 9 am and 1 pm? _41 D_
$$\begin{array}{r} 129 \\ -\ 88 \\ \hline 41 \end{array}$$

11. You need to stack some heavy boxes in a storage area. The ceiling in the storage area is 144" high. You must have a 12" clearance between the top box and the ceiling. Can you stack all three boxes pictured in *Figure 6* and still have the clearance required? _yes_

How much clearance will you have? ___14"___
$$\begin{array}{r} 48'' \\ 43'' \\ 39'' \\ \hline 130'' \end{array}$$

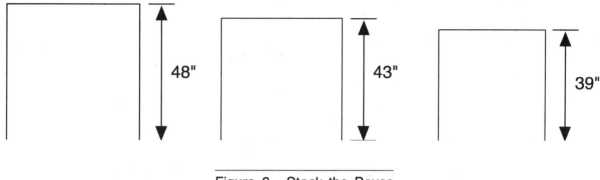

Figure 6. Stack the Boxes

1.1.11 Using The Calculator To Multiply Whole Numbers

Multiplying with a calculator is as easy as adding and subtracting with one. Here are the steps:

HOW TO MULTIPLY WHOLE NUMBERS USING A CALCULATOR

Let's use the problem 6 × 5 to practice. *30*

Step 1.   Turn the calculator ON. A zero (0) appears in the display.

Step 2.   Press 6. A 6 appears in the display.

Step 3.   Press the × key. The 6 is still displayed.

Step 4.   Press the 5 key. A 5 is displayed.

Step 5.   Press the = key. The product, 30, appears in the display.

Step 6.   Press the ON/C key to clear the calculator.

Use your calculator to figure these.

1.　　452
　　　× 4
　　　_1808_

2.　　62
　　　× 5
　　　_310_

3.　　323
　　　× 8
　　　_2584_

　　　52
4.　1,254
　　× 57
　　_9378_
　　_6270_
　　_72,078_

5.　　943
　　× 93
　　_2829_
　　_8487_
　　_87,699_

6.　4,593
　　× 47
　　_32081_
　　_18372_
　　_215801_

7.　A machine produces 465 screws in one hour. How many will it produce in 16 hours?

　　　_7,440_

　　　　　　465
　　　　　× 16
　　　　　_2790_
　　　　　_465_
　　　　　_7440_

8.　To find the area of a room, multiply its length times its width. What is the area of the room shown in *Figure 7*? _286'_

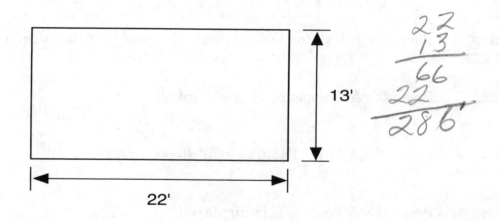

　　22
　　13
　　_66_
　　22
　　_286'_

Figure 7.　What is the Area of This Room?

9.　There are 12 inches to a foot. How many inches are there in 18 feet? _216"_

　　　18
　　× 12
　　_36_
　　18
　　_216"_

1.1.12 Using The Calculator To Divide Whole Numbers

Dividing with a calculator is as easy as the other operations. Here are the steps:

HOW TO DIVIDE WHOLE NUMBERS USING A CALCULATOR

Let's use $12 \div 4$ to practice. *3*

Step 1.   Turn the calculator ON. A zero (0) appears in the display.

Step 2.   Press the 1 key and then the 2 key. A 12 appears in the display.

Step 3.   Press the $\div$ key. The 12 is still displayed.

Step 4.   Press the 4 key. A 4 is displayed.

Step 5.   Press the = key. The quotient, 3, appears in the display.

Step 6.   Press the ON/C key to clear the calculator.

1.1.13 Expressing A Remainder As A Whole Number

When one number does not go into another number evenly, you are left with a remainder.
For example, use your calculator to figure the following problem:

USING A CALCULATOR TO EXPRESS A REMAINDER AS A WHOLE NUMBER

Problem:

There is a piece of wood 6' long. How many 4' pieces can the
worker cut from it? How many feet will he have left over?

CAN GET =1 piece (WILL HAVE 2' LEFT OVER)

Step 1.   Turn the calculator ON. A zero (0) appears in the display.

Step 2.   Press 6. A 6 appears in the display.

Step 3.   Press the $\div$ key. The 6 is still displayed.

Step 4.   Press the 4 key. A 4 is displayed.

Step 5.   Press the = key.

Step 6.   The total, 1.5, appears in the display.

Step 7.   Press the ON/C key to clear the calculator.

Step 8.   To express the ".5" as a whole number, multiply it by the number you divided by (4).   The remainder expressed as a whole number is 2' *(Figure 8)*.

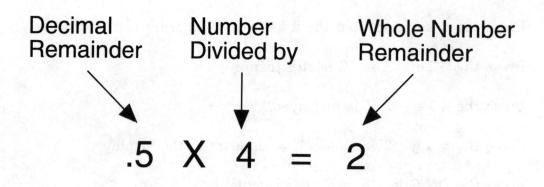

Figure 8.   From Decimal Remainder to Whole Number Remainder

The answer to the problem, "How many 4' pieces can the worker cut from a 6' piece of wood? How many feet will he have left over?" is 1 piece with 2' left over.

SELF-CHECK REVIEW / PRACTICE QUESTIONS 12

Use your calculator to figure these.   Express any remainders as whole numbers.

1.   15 ÷ 3 = ___5___   2.   36 ÷ 4 = ___9___   3.   54 ÷ 9 = __6__

4.   17 ÷ 6 = __2.86__   5.   39 ÷ 7 = __5.57__   6.   68 ÷ 8 = __8.5__

7.   You have a 100 gallon tank filled with liquid.   You need to put the liquid into 20 gallon containers.   How many 20 gallon containers can you fill? ___5___

8. An assembly requires 5 bolts. You have 127 bolts available. How many assemblies can you complete? __*25*__   How many bolts (in whole numbers) will you have left over? __*2*__

    127 ÷ 5 = 25
    2

9. A wooden crate of parts weighs 1,090 pounds. The crate itself weighs 10 pounds. There are 60 parts in the crate. How much does each part weigh? __*18*__

    1080 ÷ 60 = 18

    1090
    - 10
    1,080

10. You earn 1 hour of vacation time for every 40 hours worked. In a given time period, you worked 1,970 hours. How many hours of vacation did you earn? __*49.15 min.*__

    1970 ÷ 40
    49 HRS 15 MINUTES

2.0.0 INTRODUCTION TO FRACTIONS AND MEASUREMENT

In the construction trade, you will be asked to use a ruler to measure various objects. You will also be expected to perform operations with these measurements.

There are two types of rulers you may see on the job: The **Standard English ruler** (standard ruler) *(Figure 9)* or the **Metric ruler** *(Figure 10)*.

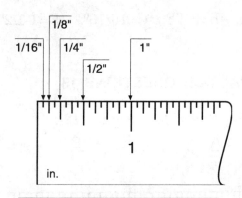

Figure 9. The Standard Ruler

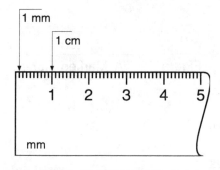

Figure 10. The Metric Ruler

In this section, we are working only with the Standard ruler. Later we will work with the Metric ruler. The Standard ruler is divided into whole inches and then halves, fourths, eighths, and sixteenths. Some are divided into thirty-seconds, and sometimes, sixty-fourths. These represent fractions of an inch. In this unit, you will be working with a Standard ruler and fractions to solve problems.

2.1.0   USING THE STANDARD RULER *(Figure 11)*

Observe the following distances from the start of the ruler:

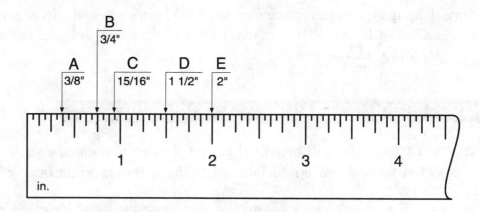

Figure 11. Markings on a Standard Ruler

A=3/8";   B=3/4";   C=15/16";   D=1-1/2";   E=2"

SELF-CHECK REVIEW / PRACTICE QUESTIONS 13

Fill in the blanks below using this ruler *(Figure 12)*:

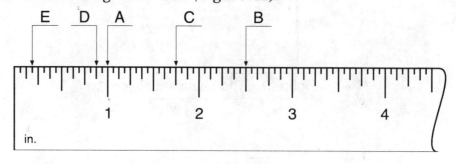

Figure 12.   A Marked Ruler

1.   A is at the_1_" mark.

2.   B is at the 2-_1_/2" mark.

3.   C is at the 1-_3_/4" mark.

4.   D is at the _7_/8" mark.

5.   E is at the _2_/16" mark.

6.   Use the Standard ruler to determine the length of the line below to the nearest 1/4th of an inch.

What is the length of the line? _____

_____

7.   Measure this line to the nearest 16th of an inch.

_____

What is the length of the line? _____

Draw lines of the following lengths:

8.   3-1/8"

Draw here:

9.   4-5/16"

Draw here:

10.   2-1/4"

Draw here:

2.1.1   What Are Fractions?

A **fraction** divides whole units into parts. Common fractions are written as two numbers, separated by a slash or by a horizontal line, as such:

1/2   or $\frac{1}{2}$

The slash or horizontal line means the same thing as the ÷ sign. So think of a fraction as a division problem. The fraction 1/2 means 1 divided into 2 equal parts and is read "one half."

The bottom number of the fraction is called the **denominator**. It tells you the number of parts by which the whole unit is being divided. The top number is called the **numerator**. It tells you how many parts are being used.

In the fraction 1/2, the 1 is the numerator; the 2 is the denominator.

What measurement is the arrow in *Figure 13* pointing to?

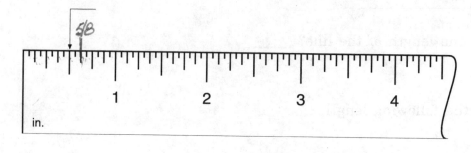

Figure 13. What's the Measurement?

Circle the correct answer.

A. 8/16   B. 2/4   C. 4/8   D. 1/2   E. ALL are correct!

The correct answer is: E ... all of the answers are correct! Let's find out why.

2.1.2 Finding Equivalent Fractions

You can see that 1/2" = 2/4" = 4/8" = 8/16". These fractions are called **equivalent fractions**.

Equivalent means that they have the same value - they are equal. If you cut off a piece of wood 8/16" long, and I cut off a piece 1/2" long, we would have the same length of wood.

When you measure objects, you often need to record all measurements as common fractions such as in sixteenths of an inch. In this way you may easily compare, add, and subtract fractional measurements. For this reason, you need to know how to find equivalent fractions.

HOW TO FIND EQUIVALENT FRACTIONS

1/2" equals how many sixteenths of an inch?

8/16

Step 1.   To find equivalent fractions, multiply the numerator and the denominator by the same number.

For example: Multiply 1/2" to calculate sixteenths of an inch.

$$\frac{1 \times 8 = 8}{2 \times 8 = 16}$$

Answer: 1/2" is equivalent to 8/16"

SELF-CHECK REVIEW / PRACTICE QUESTIONS 14

What are the equivalents of these measurements?

REPASAR ESTA PARTE

1.   1/4" = _4_/16"   2.   2/16" = _4_/32"
3.   3/4" = _3_/8"   4.   1/2" = _2_/4"   *3/3 = 12*
5.   1/8" = _2_/16"   6.   3/8" = _12_/32"
7.   1/2" = _4_/8"   8.   3/4" = _48_/64"   *2/3 = 8*
9.   2/3" = _8_/12"   10.   1/2" = _8_/16"
                                                          1/3 = 4

2.1.3   Reducing Fractions To Their Lowest Form

If you find that the measurement of something is 4/16, you may want to reduce it to its lowest form so the number is easier to work with. *= 1/4"*

HOW TO REDUCE FRACTIONS TO THEIR LOWEST FORM

To find the lowest form of 4/16, you will use division.

Step 1.   Divide the numerator and the denominator by the same number.

To reduce a fraction, ask yourself: What is the largest number that I can divide evenly into both the numerator and denominator? If there is no number (other than 1) which will divide evenly into both numbers, the fraction is already in it lowest form.

> *TIP*   If the numerator and denominator are both even,
> you know that you can divide each by 2.

In this example, you could divide the numerator and denominator by 4.

$$\frac{4 \div 4 = 1}{16 \div 4 = 4}$$

SELF-CHECK REVIEW / PRACTICE QUESTIONS 15

Find the lowest form of each fraction.

1.   4/16 = 1/4   2.   2/4 = 1/4

3.   2/8 = 1/4   4.   6/12 = 1/2

5.   12/32 = 3/8   6.   2/16 = 1/8

7.   4/8 = 1/2   8.   9/12 = 3/4

9.   4/64 = 1/16   10.   9/32 = 3/8

2.1.4   Comparing Fractions

Which measurement is larger: 2/3 or 5/8? Would you have more pizza if you had 2 pieces from a pie that was cut up in 3 equal slices *(Figure 14)?*

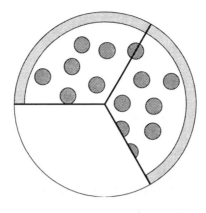

Figure 14. 2/3 of a Pizza

Or would you have more pizza if you had 5 pieces of the same size pie that had been cut up in eighths *(Figure 15)?*

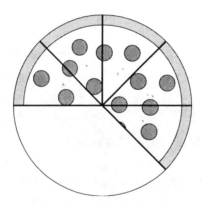

Figure 15. 5/8 of a Pizza

As you can see, it's hard to compare fractions that do not have common denominators, just as it's hard to compare pizzas that are cut up differently.

HOW TO FIND AND CONVERT TO COMMON DENOMINATORS

Using our pizza pie, remember that we are trying to determine which amount of pie is larger:

$$\frac{2}{3} \quad \text{or} \quad \frac{5}{8}$$

To compare, you need to find the common denominator of the pizza slices. The common denominator is a number that both denominators can go into evenly.

Step 1.   Multiply the two denominators together ($3 \times 8 = 24$). This is a common denominator between the two fractions.

You found the common denominator so that you can more easily compare the pieces. Now we'll **convert** the two fractions so that they will have the same denominator:

Step 2.   Convert each to fractions having this common denominator.

$$\frac{2}{3} \times \frac{8}{8} = \frac{16}{24}$$

$$\frac{5}{8} \times \frac{3}{3} = \frac{15}{24}$$

Now it's easy to compare the two fractions to see which is larger. You'd have more pizza if you chose 2/3 (you'd have 16/24 as opposed to 15/24 of the pizza).

SELF-CHECK REVIEW / PRACTICE QUESTIONS 16

Find a common denominator for these pairs of fractions:

1. $\frac{2}{3},\ \frac{5}{6}$ _18_

2. $\frac{1}{4},\ \frac{3}{5}$ _20_

3. $\frac{5}{6},\ \frac{3}{7}$ _42_

4. $\frac{4}{9},\ \frac{3}{8}$ _72_

5. $\frac{8}{12},\ \frac{2}{3}$ _24_

6. $\frac{3}{9},\ \frac{5}{7}$ _63*_

7. $\frac{4}{11},\ \frac{5}{8}$ _88_

8. $\frac{7}{32},\ \frac{5}{16}$ _64_

9. $\frac{5}{6},\ \frac{5}{12}$ _24,12_

2.1.5   Adding Fractions

How many *total slices* would you have if you were given 2/3 of one pizza plus 5/8 of a second pizza of the same size?   To answer this question, you will have to add two fractions.

HOW TO ADD FRACTIONS

Step 1.   Find the common denominator of the fractions you wish to add.

The common denominator of 2/3 and 5/8 is 24.

Step 2.   Convert the fractions to equivalent fractions with the same denominator.

This is how to convert the fractions to equivalent fractions:

$$\frac{2}{3} \quad \times \quad \frac{8}{8} \quad = \quad \frac{16}{24}$$

$$\frac{5}{8} \quad \times \quad \frac{3}{3} \quad = \quad \frac{15}{24}$$

Step 3.   Add the numerators of the fractions.   Place this sum over the denominator.

Now you can (3) add the numerators of the fractions and place this sum over the denominator.

$$\frac{16}{24} \quad + \quad \frac{15}{24} \quad = \quad \frac{31}{24}$$

Step 4.   Reduce the fraction to its lowest terms.   This means that there is no number other than 1 that can go evenly into both the numerator and the denominator or that the fraction is not an **improper fraction** (meaning the numerator is not larger than the denominator).

The answer to "How many total slices of pizza will you have if you add 2/3 plus 5/8 pizzas together?" is 31/24 total slices.

In this case, you would need to reduce the fraction 31/24 to its lowest terms, as this fraction *is* an improper fraction - the numerator is larger than the denominator. We will soon learn how to convert improper fractions to **mixed numbers**.   But for now, try these.

Add these. Reduce the sum to lowest terms.

1. $\dfrac{1}{8}$ + $\dfrac{4}{16}$ = _3/8___

2. $\dfrac{2}{9}$ + $\dfrac{6}{12}$ = _____

3. $\dfrac{2}{5}$ + $\dfrac{1}{4}$ = _____

4. $\dfrac{3}{4}$ + $\dfrac{2}{3}$ = _1-5/12___

5. $\dfrac{1}{3}$ + $\dfrac{1}{15}$ = _____

6. $\dfrac{5}{6}$ + $\dfrac{7}{8}$ = _____

7. $\dfrac{3}{4}$ + $\dfrac{5}{6}$ = _____

8. $\dfrac{3}{8}$ + $\dfrac{5}{6}$ = _____

9. $\dfrac{5}{12}$ + $\dfrac{1}{3}$ = _____

HOW TO CONVERT IMPROPER FRACTIONS TO MIXED NUMBERS

Improper Fractions are fractions where the numerator is larger than the denominator. Usually, you will want to convert Improper Fractions to mixed numbers. For this exercise, we will convert the improper fraction 31/24, the answer to the pizza problem, to a mixed number.

Step 1.   Divide the numerator by the denominator.

31/24 is actually (31 ÷ 24).

Step 2.   Place the remainder over the denominator

31 ÷ 24 = 1 with a remainder of 7.

Write this as the mixed number 1-7/24.

Step 3.   Reduce the fraction part of the mixed number to its lowest terms, if necessary. (7/24 is in its lowest terms, as there is no number that can go into both 7 and 24).

The answer is 1-7/24.

Convert these improper fractions to mixed numbers.

1. $\dfrac{13}{10}$ =   *1-3/10*

2. $\dfrac{27}{8}$ =   *3-3/8*

3. $\dfrac{59}{10}$ =   *5-9/10*

4. $\dfrac{27}{5}$ =   *5-2/5*

5. $\dfrac{46}{3}$ =   *15 1/3*

6. $\dfrac{64}{7}$ =   *9-1/7*

7. $\dfrac{76}{5}$ =   *13-1/5*

8. $\dfrac{92}{12}$ =   *8-12/7*

9. $\dfrac{108}{16}$ =   *6-3/4 ; 6-12/16*

2.1.6 Subtracting Fractions

Subtracting fractions is very much like adding fractions. You must find a common denominator before subtracting.

Problem:

You have a piece of wood 7/8 of a foot long. You use 1/4 of a foot. How much will you have left?

$\dfrac{7}{8}$ $\dfrac{7}{8}$

1. Determine the common denominator. In this case it is 8.

$\dfrac{1}{4}$ $\times$ $\dfrac{2}{2}$ = $\dfrac{2}{8}$

2. Multiply the entire fraction (1/4) by 2 to get the common denominator of 8.

$\dfrac{7}{8}$ − $\dfrac{2}{8}$ = $\dfrac{5}{8}$ of a foot is left

3. Subtract the numerators. Reduce if possible. 5/8 of a foot is left.

Subtract these. Reduce the difference to lowest terms.

1. $\dfrac{3}{8} - \dfrac{5}{16} =$ _____ 1/16

2. $\dfrac{8}{12} - \dfrac{5}{9} =$ _____ 1/9

3. $\dfrac{3}{4} - \dfrac{2}{5} =$ _____ 7/20

4. $\dfrac{11}{12} - \dfrac{4}{5} =$ _____ 7/60

5. $\dfrac{2}{3} - \dfrac{5}{8} =$ _____ 1/24

6. $\dfrac{9}{10} - \dfrac{3}{4} =$ _____ 3/20

7. $\dfrac{11}{16} - \dfrac{1}{2} =$ _____ 3/16

8. $\dfrac{7}{8} - \dfrac{1}{3} =$ _____ 13/24

9. $\dfrac{4}{5} - \dfrac{1}{3} =$ _____ 7/15

Sometimes you must subtract a fraction from a whole number. For example:

Problem:

You need to take 1/4 of a day off from a 5 day workweek. How many days will you be working that week?

Here is how to set up this type of problem:

HOW TO SUBTRACT A FRACTION FROM A WHOLE NUMBER

Step 1. To subtract a fraction from a whole number, borrow 1 from the whole number to make into a fraction.

$$
\begin{array}{c}
5 \\
-1/4
\end{array}
\quad = \quad
\begin{array}{c}
4 + 1 \\
-1/4
\end{array}
$$

Step 2. convert the 1 to a fraction having the same denominator as the number you are subtracting.

$$
\begin{array}{c}
5 \\
-1/4
\end{array}
\quad = \quad
\begin{array}{c}
4 + 4/4 \\
-1/4
\end{array}
$$

Step 3.   Subtract and reduce to lowest terms.

$$
\begin{array}{rcl}
5 & = & 4 + 4/4 \\
- \ 1/4 & & \underline{- \ 1/4} \\
& & 4 \ - \ 3/4 \ \text{days}
\end{array}
$$

SELF-CHECK REVIEW / PRACTICE QUESTIONS 20

Solve these.   Reduce to lowest terms.

1.   8   7-1/4
      $\underline{- \ 3/4}$

2.   12   11-3/8
      $\underline{- \ 5/8}$

3.   8 1/5   5 - 33/40
      $\underline{- \ 2 \ 3/8}$

4.   A motor with four pieces of steel stacked under its base is found to be too high. The total thickness of these pieces is 5/64". It is necessary to remove a piece 1/32" thick. What is the total thickness of the steel left under the motor base, reduced to lowest terms? ___3/64"___

5.   If two punches, one 4-1/64" long and the other 4-3/32" long, are made from a bar of stock 9-7/16" long, what length of stock is not used? __1 - 21/64"__

6.   A 29/32" diameter hole must be enlarged to 59/64" diameter to insert a nail. How much larger than the original hole will this be?__1/64"__

7.   A rough opening for a window measures 36-3/8". A window to be placed in the rough opening measures 35-15/16". How much total clearing will exist between the window and the rough opening? __7/16__

8.   You are cutting four pieces of electrical conduit. The lengths of conduit are:

      6-1/2'   5-3/4'   1-1/3'   3-1/4'

   What is the total length of the cut conduit, reduced to its lowest form? _____

9. The specified dimension of a pushrod is 7-7/16" ± 1/32". (See note below.) What is the minimum and maximum dimension? State your answers as mixed numbers.

Minimum dimension: _____   Maximum dimension: _____

Note   The ± sign indicates that the pushrod measurement has a **tolerance** of 1/32". This means that the measurement can be off either 1/32" LESS than the specified dimension (7-7/16") or 1/32" MORE than the specified dimension.

10. For a job, a mason estimates that 18-cubic yards, 9-1/3 cubic yards, 12-1/2 cubic yards, and 17-2/3 cubic yards of mortar are needed. What is the total amount of mortar required for this job?

2.1.7   Multiplying Fractions

Multiplying and dividing fractions is very different from adding and subtracting fractions. You do not have to find a common denominator when you multiply or divide fractions.

In a word problem, multiplication is usually indicated by the word 'OF'. If a problem asks "What is 2/3 of 9?" then write the problem as 2/3 × 9/1.

HOW TO MULTIPLY FRACTIONS

Using 4/5 × 5/6 as an example:

Step 1. Multiply the numerators together to get a new numerator. Multiply the denominators together to get a new denominator.

$$\frac{4}{5} \times \frac{5}{6} = \frac{20}{30}$$

Step 2. Reduce if possible (20/30 reduces to 2/3).

Solve these. Reduce if possible.

1. $\dfrac{4}{5} \times \dfrac{5}{8} =$ _____   2. $\dfrac{9}{10} \times \dfrac{5}{12} =$ _____   3. $\dfrac{4}{7} \times \dfrac{3}{16} =$ _____

4. $\dfrac{3}{4} \times \dfrac{7}{8} =$ _____   5. $\dfrac{3}{16} \times \dfrac{2}{3} =$ _____   6. $\dfrac{4}{9} \times \dfrac{3}{16} =$ _____

7. What is 2/3 of 15? _____   8. What is 5/16 of 12? _____

2.1.8   Dividing Fractions

Dividing fractions is very much like multiplying fractions, with one difference. You must **invert**, or flip, the fraction you are dividing by.

HOW TO DIVIDE FRACTIONS

Using $1/2 \div 3/4$ as an example:

Step 1. Invert the fraction you are dividing by (3/4).

$\dfrac{3}{4}$ becomes $\dfrac{4}{3}$

Step 2. Change the division sign ($\div$) to a multiplication sign ($\times$).

$$\dfrac{1}{2} \div \dfrac{3}{4} = \dfrac{1}{2} \times \dfrac{4}{3}$$

Step 3. Multiply the fraction as instructed earlier.

$$\dfrac{1}{2} \times \dfrac{4}{3} = \dfrac{4}{6}$$

Step 4.   Reduce if possible.

> $\frac{4}{6}$ reduces to $\frac{2}{3}$

Note   If you are working with a mixed number (for example, 2-1/3), you must convert it to a fraction before inverting. Do this by multiplying the denominator (3) by the whole number (2) and add the numerator (1) and place the result over the denominator (3). It looks like:

>   2-1/3   =   (3 × 2) + 1 = $\frac{7}{3}$

When dividing by a whole number, place the whole number over 1 and then invert it.

>   For example: 1/2 ÷ 4   =
>
>   1/2 ÷ 4/1   =
>
>   1/2 × 1/4   = 1/8

SELF-CHECK REVIEW / PRACTICE QUESTIONS 22

Divide. Reduce answers to lowest terms (no improper fractions allowed here!)

1. $\frac{3}{8}$ inch ÷ 3   =   _____

2. $\frac{5}{8}$ inch ÷ $\frac{1}{2}$ =   *1-1/4*

3. $\frac{3}{4}$ inch ÷ $\frac{3}{8}$ =   _____

4.   5 $\frac{1}{2}$ yard ÷ $\frac{1}{4}$ =   _____

5. On a scale drawing, 1/4" represents 1'. How many feet are represented by 8-1/2"?
    _____

6. How many 7/8" lengths can be cut from a 7" strip? _____

7. If 7-3/4 dozen of connectors cost $31, what is the cost per dozen? _____

3.0.0   WORKING WITH DECIMALS AND PERCENTS

Decimals represent values less than one whole unit. You are already familiar with decimals in the form of money.

    25¢ = 0.25   or   25/100

    10¢ = 0.10   or   10/100

    50¢ = 0.50   or   50/100

On the job, you may need to use decimals to read instruments or calculate flow rates. Look at the scale *(Figure 16)* on a typical machinist's rule.

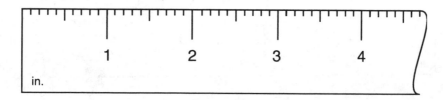

Figure 16. The Machinist's Rule (Divided into Tenths)

Each number shows the distance, in inches, from the squared end of the rule. The marks between the numbers divide each inch into ten equal parts. Each of these ten parts are referred to as **tenths**.

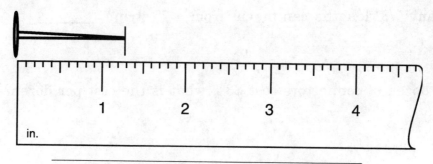

Figure 17. Showing 1.3" on a Machinist's Rule

This nail spans one whole inch plus three tenths of a second inch *(Figure 17)*. It is one and three tenths of an inch long. This is written as 1.3".

SELF-CHECK REVIEW / PRACTICE QUESTIONS 23

What are the decimals measurements for these indications on the rule *(Figure 18)?* Write your answer in the spaces indicated below.

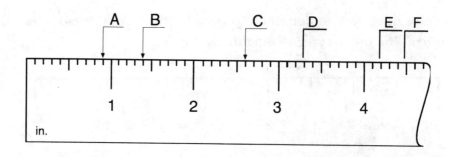

Figure 18. Markings on a Machinist's Rule

A = _____   B = _____   C = _____

D = _____   E = _____   F = _____

The illustration below *(Figure 19)* compares whole number place values with decimal place values.

WHOLE NUMBERS			DECIMALS	
1	ones			
10	tens		.1	tenths
100	hundreds		.01	hundredths
1000	thousands		.001	thousandths
10,000	ten-thousands		.0001	ten-thousandths
100,000	hundred-thousands		.00001	hundred-thousandths
1,000,000	millions		.000001	millionths

Figure 19. Comparing Whole Number Place Values to Decimal Place Values

To read a decimal, say the number as it is written and then the name of its place value. For example:

0.56 is read as "fifty-six hundredths"

0.560 is read as "five hundred sixty thousandths"

0.056 is read as "fifty-six thousandths"

Mixed numbers also appear in decimals. 15.7 is read as "fifteen and seven tenths." Notice the use of the word "and" to separate the whole number from the decimal.

SELF-CHECK REVIEW / PRACTICE QUESTIONS 24

Write the decimals as words and the words as decimals.

1. 0.4 = _____

2. 0.5 = _____

3. 0.12 = _____

4. 0.25 = _____

5. 0.245 = _____

6. 2.5 = _____

7. 6.12 = _____

8. 5.025 = _____

9. four thousandths = _____

10. eighteen hundredths = _____

3.1.0 COMPARING DECIMALS

Which decimal is the larger of the two?

    0.4 or 0.42 ?

Here's how to compare decimals:

HOW TO COMPARE DECIMALS

Step 1.   Line up the decimal points of all the numbers.

        0.4

        0.42

Step 2.   Place zeros to the right of each number until all numbers end with the same place value.

        0.40

        0.42

Step 3.   Compare the numbers.

        0.42 (42 hundredths) is larger than .40 (40 hundredths).

SELF-CHECK REVIEW / PRACTICE QUESTIONS 25

Put these decimals in order from *smallest* to *largest* (write 1 (smallest) thru 4 (largest) beneath the decimals).

1.   0.400   0.004   0.044   0.404

     _____   _____   _____   _____

2.   0.567   0.059   0.56   0.508

     _____   _____   _____   _____

3.   0.3200   0.0320   0.3020   0.0032

     _____   _____   _____   _____

4.   0.867   0.086   0.0086   0.870

     _____   _____   _____   _____

3.1.1 Adding And Subtracting Decimals

How to Add and Subtract Decimals

There is only one major rule to remember when adding and subtracting decimals:

Step 1.   KEEP YOUR DECIMAL POINTS LINED UP !!!

Suppose you want to add 4.76 and 0.834. Line up the problem like this:

```
  4.760   You can add a 0 to help keep the numbers lined up.
+ 0.834
  5.594
```

The same thing is true for subtraction of decimals. Line up the decimal points.

```
  5.6        5.600        Notice that two zeros were added to the end of
- 2.724    - 2.724        the first number to make it easier to borrow
  2.876                   from.
```

SELF-CHECK REVIEW / PRACTICE QUESTIONS 26

Try these addition and subtraction problems. Do not forget to line up the columns.

1.
```
    2.00
    4.00
+   5.00
```

2.
```
    1.50
    2.30
+   1.80
```

3.
```
    3.40
    2.30
+   1.68
```

4. $1.82 + 3.41 + 5.25 =$ _____

5. $6.43 + 86.4 =$ _____

6.   Find the total thickness of a piece of sheet metal 0.0784" thick and a piece of band iron 0.25" thick. _____

7. Yesterday, a wood yard contained 6.7 tons of wood. Since then, 2.3 tons have been removed. How much wood remains? _____

8. A dump truck has a load limit of 13 tons. You place 11.79 tons of sand in the truck. How many more tons of sand can be loaded? _____

3.1.2 Multiplying Decimals

While unloading wood panels, you measure one panel to be 4.5' in width. You have 7 panels. What is the total width if all the panels were laid side-by-side?

HOW TO MULTIPLY DECIMALS

Step 1.   Set up the problem just like the multiplication of whole numbers.

$$\begin{array}{r} 4.5 \\ \times\ 7 \\ \hline \end{array}$$

Step 2.   Proceed to multiply.

$$\begin{array}{r} 4.5 \\ \times\ 7 \\ \hline 315 \end{array}$$

Step 3.   Once you have the product of the number, count the number of digits to the right of the decimal point in both numbers being multiplied. (There is only one decimal point (4.5) and only one number to the right of it.)

Step 4.   In the product, count over the same number of digits (from right to left) and place the decimal point there.

$$\begin{array}{r} 4.5 \\ \times\ 7 \\ \hline 31.5 \end{array}$$ (count 1 total digit to the right of the decimal point in the two numbers)

(count in 1 digit from right to left in the product)

Note   You may have to add a zero if there are not as many digits as there are total number of numbers to the right of the decimal points of the two numbers.

$$0.507 \quad \text{(count 6 total digits to the right of the}$$
$$\underline{\times\ 0.022} \quad \text{decimal point in the two numbers)}$$
$$1014$$
$$1014$$
$$\underline{000\quad}$$
$$11154 = .011154 \quad \text{(count in 6 digits from right to left in the product. Here, you'll need to add a 0)}$$

SELF-CHECK REVIEW / PRACTICE QUESTIONS 27

You are machining a part. The starting thickness of the part is 6.086 inches. You take three cuts. Each cut is three-tenths of an inch.

1.   How much material is removed?_____

2.   What is the remaining thickness of the part? _____

An electrician wants to know if a light circuit is overloaded. The circuit supplies two different machines. The first machine has 11 bulbs lit. Each bulb uses 4.68 watts. The second machine has 7 bulbs lit. Each of these bulbs uses 5.14 watts.

3.   How many watts are needed by the lights on the first machine?   _____

4.   How many watts are needed by the lights on the second machine?   _____

5.   How many watts are needed by the lights on both machines?   _____

6.   The circuit has a rating of 95.5 watts. Is it overloaded?   _____

7.   Ceramic tile weighs 4.75 pounds per square foot. Find the weight of 128 square feet of ceramic tile. _____

Find the totals and the grand total for the items on the chart below.

	ITEM	QUANTITY	UNIT COST	TOTAL
8.	A	5	$2.45	_____
9.	B	7	$3.75	_____
10.	C	6	$0.95	_____
11.			GRAND TOTAL:	_____

3.1.3 Dividing Decimals

When would you have to divide with decimals? Here's an example:

Problem:

You need to cut a 44.22" pipe into as many 22" pieces as possible. How many 22" pieces will you be able to cut?

There are three types of division problems.

1. Those that have a decimal point in the number being divided (the dividend),

    $22 \overline{)\, 44.22}$ (this is the problem presented above)

2. Those that have a decimal point in the number you are dividing by (the divisor), and

    $0.22 \overline{)\, 4,422}$

3. Those that have decimal points in both numbers (the divisor and the dividend).

    $0.22 \overline{)\, 44.22}$

Let's look at each type of problem.

HOW TO DIVIDE WITH A DECIMAL POINT IN THE DIVIDEND

We'll use 44.22 ÷ 22 for this exercise. (This is the problem presented above).

Step 1.   Place a decimal point directly above the decimal point in the dividend.

$$
22 \overline{)\ 44.22}
$$

Step 2.   Divide as usual.

```
        2.01
22 ) 44.22
     44.
       22
       22
       00
```

How many 22" pieces of pipe will you have? Two, with a little left over.

SELF-CHECK REVIEW / PRACTICE QUESTIONS 28

Divide these.   Don't go any further than the hundredths (.01) place in your answer.

1.   45.36 ÷ 18 =   2.   4.536 ÷ 18 =   3.   .4536 ÷ 18 =

4.   25) 10.20   5.   6) 31.2   6.   98) 2.156

HOW TO DIVIDE WITH A DECIMAL POINT IN THE DIVISOR

We'll use 4,422 ÷ .22 for this exercise.

Step 1.   Move the decimal point in the number you are dividing by (the divisor) to the right until you have a whole number.

$$.22.\overline{)\,4{,}422.}$$

Step 2.   Move the decimal point in the number you are dividing into (the dividend) the same number of places to the right. You may have to add zeros first. Divide as usual.

```
         20100.
   22 ) 4422.00.
        44
        02
        00
        022
         22
         00
         00
          00
```

SELF-CHECK REVIEW / PRACTICE QUESTIONS 29

Divide these. Don't go any further than the hundredths (.01) place in your answer.

1.   282 ÷ 14.1 =

2.   694 ÷ 3.2 =

3.   99 ÷ .45 =

4.   .25)‾ 1,020

5.   0.6)‾ 312

6.   0.98)‾ 2,156

HOW TO DIVIDE WITH A DECIMAL POINT IN THE DIVIDEND AND DIVISOR

We'll use 44.22 ÷ .22 for this exercise.

Step 1.   Move the decimal point in the number you are dividing by (the divisor) to the right until you have a whole number.

$$.22. \overline{)\,44.22}$$

Step 2.   Move the decimal point in the number you are dividing into (the dividend) the same number of places to the right.   Divide as usual.

```
          201
    22 )  44.22.
          44
          02
          00
          22
          22
           0
```

SELF-CHECK REVIEW / PRACTICE QUESTIONS 30

Divide these.   Don't go any further than the hundredths (.01) place in your answer.

1.   20.82 ÷ 4.24 = _____   2.   38.9 ÷ 3.7 = _____   3.   9.9 ÷ .45 = _____

4. $.25 \overline{)\,10.20}$ 5. $0.6 \overline{)\,31.2}$ 6. $0.98 \overline{)\,2.156}$

3.1.4 Rounding Decimals

Sometimes the answer is a bit more precise than you require. For example:

> *Problem*:
>
> You need to cut a 107.5" pipe into as many 4.25" pieces as possible. How many 4.25" pieces will you be able to cut?

The precise answer is 25.29411764. But you might only need to measure it to the nearest tenth. What would you do?

HOW TO ROUND DECIMALS

For this exercise, round 25.29411764 to the nearest tenth (.1).

Step 1. Underline the place to which you are rounding.

25.<u>2</u>9411764

Step 2. Look at the digit one place to its right.

25.<u>29</u>411764

Step 3. If the digit to the right is 5 or more, you will round up by adding 1 to the underlined digit. If the digit is less than 5, leave the underlined digit the same.

25.<u>39</u>411764

Step 4. Drop all other digits to the right.

25.3

1.   You need to cut a 107.5" pipe into as many 4.25" pieces as possible. How many 4.25"
        pieces will you be able to cut? _____

Find the miles per gallon (MPG) for the following vehicles. Round to the nearest tenth.

	CAR	MILES	GALS USED	MPG
2.	A	582	38.6	____
3.	B	650	42.4	____
4.	C	620	41.7	____
5.	D	622	40.1	____

6.   Which car (A, B, C, or D) got the best mileage (MPG)? _____

You are asked to purchase the lowest-cost wire. You are comparing two companies' prices.
What is the cost per pound of each? Round to the nearest tenth.

	COMPANY	TOTAL COST	TOTAL WEIGHT	COST PER LB.
7.	A	$114.95	28.34 lbs.	_____
8.	B	$ 95.79	26.85 lbs.	_____

9.   Which company will you buy the wire from? _____

3.1.5   Using The Calculator To Add, Subtract, Multiply, And Divide Decimals

Performing operations on the calulator using decimals is very much like performing the
operations on whole numbers.

HOW TO ADD, SUBTRACT, MULTIPLY, OR DIVIDE DECIMALS WITH A CALCULATOR

We'll use: 45.6 + 5.7

Step 1.   Turn the calculator ON. A zero (0) appears in the display.

Step 2.   Press 45.6. A 45.6 appears in the display.

Step 3.   Press the + key. The 45.6 is still displayed.

Note   For this step, press whichever operation key the problem calls for:

+ to add;   – to subtract;   × to multiply;   ÷ to divide.

Step 4.   Press 5.7. A 5.7 is displayed.

Step 5.   Press the = key. The sum (add), difference (subtract), product (multiply), or quotient (divide) will appear on your display.

$$45.6 + 5.7 = 51.3 \qquad\qquad 45.6 - 5.7 = 39.9$$
$$45.6 \times 5.7 = 259.92 \qquad\qquad 45.6 \div 5.7 = 8$$

Step 6.   Press the ON/C key to clear the calculator.

SELF-CHECK REVIEW / PRACTICE QUESTIONS 32

Do these on your calculator. Round all to nearest hundredth (.01). OBSERVE THE OPERATIONAL SIGNS: +   ×   –   ÷

1.   45.89
     +7.85

2.   7.6
     × .12

3.   685.79
     – 56.266

4. $6.45 \div 3.25 =$ _____

5.   34.76
     +3.64

6.   8.2
     × .74

7.   553.75
     – 74.536

8. $7.47 \div 0.24 =$ _____

9. 75.94   10. 5.9   11. 663.42   12. 7.65 ÷ 2.85 = _____

    +2.74   × .64   − 91.251

3.1.6   Converting Decimals To Percents And Percents To Decimals

What are percents? When a whole number is divided into 100 parts, you can express some part of the whole as a percent. Let's look at an example.

Problem:

This tank *(Figure 20)* has a capacity of 100 gallons. It is now filled with 50 gallons. What percent of the tank is filled?

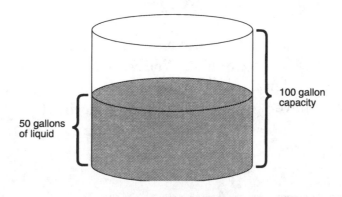

Figure 20.   100 Gallon Capacity Tank

If you answered 50%, you are correct.

Percent means **out of 100**. How many gallons out of 100 does the tank contain? It contains 50 out of 100, or 50%.

Percents are an easy way to express parts of a whole. Decimals and fractions also express parts of a whole. Let's look at the relationship between the three (percents, decimals, + fractions).

The tank above was 50% full. If you expressed this as a fraction you would say it was 1/2 full. You could express this as a decimal and say it's 0.50 full.

Sometimes you may need to express decimals as percents or percents as decimals. Suppose you are preparing a gallon of cleaning solution. The mixture should be from 10% to 15% cleaning agent. The rest should be water. You have 0.12 gallon of cleaning agent. Will you have enough to prepare a gallon of the solution?

To answer the question, you must convert a decimal (0.12) to a percent.

HOW TO CONVERT DECIMALS TO PERCENTS

We will change 0.12 to a percent for this exercise.

Step 1. Multiply the decimal by 100. (Move the decimal point 2 places to the right.)

0.12 × 100 = 12

Step 2. Add a % sign.

12%

You have enough cleaning agent to make the solution. Recall that the mixture should be from 10% to 15% cleaning agent. You have 12%.

You may also need to convert percents to decimals. Let's say that another mixture should contain 22% of a certain chemical by weight. You're making 1 pound of the mixture. You weigh the ingredients on a digital scale. How much of the chemical should you add? To answer this, you must convert a percent (22%) to a decimal.

HOW TO CONVERT PERCENTS TO DECIMALS

We will change 22% to a decimal for this exercise.

Step 1. Drop the % sign.

22

Step 2. Divide the number by 100. (Move the decimal point 2 places to the left.)

22 ÷ 100 = 0.22

The answer to the problem is that you would add 0.22 pounds of the chemical to make a 22% solution.

Express the following decimals as percents.

1.	0.62 = ____	2.	0.625 = ____	3.	0.7 = ____
4.	6.34 = ____	5.	0.12 = ____	6.	64.2 = ____

Express the following percents as decimals.

7.	72% = ____	8.	12.5% = ____	9.	200% = ____
10.	34% = ____	11.	35.5% = ____	12.	350% = ____

3.1.7   Converting Fractions To Decimals And Decimals To Fractions

You will often need to change a fraction to a decimal. For example:

Problem:

You need 3/4 of a dollar. How do you convert 3/4 to its decimal equivalent?

HOW TO CONVERT A FRACTION TO A DECIMAL

Step 1.   Divide the numerator by the denominator.

$$4 \overline{)\,3.0}$$

You need to put the decimal point and the zero after the number 3 because you need a number large enough to divide by 4.

Step 2.   Put the decimal point directly above its location within the division symbol.

$$4 \overline{)\,3.0}^{\;\;.}$$

Step 3. Once the decimal point is in its proper place above the line, you can divide as you normally would. The decimal point "holds" everything in place.

$$\begin{array}{r} .75 \\ 4\overline{)\,3.00} \\ \underline{28} \\ 20 \\ \underline{20} \end{array}$$

Step 4. Read the answer.

The fraction 3/4 converted to a decimal is 0.75.

In relation to the earlier problem: 3/4 of a dollar is the same as $0.75.

SELF-CHECK REVIEW / PRACTICE QUESTIONS 34

Convert the following fractions to decimals.

1. 1/4 = _____   2. 3/4 = _____

3. 1/8 = _____   4. 5/16 = _____

5. 20/64 = _____   6. 6/32 = _____

Convert the following inches to their decimal equivalents in feet. For example:

Problem:

3" = what decimal equivalent in feet?

HINT First express the inches as a fraction having 12 as the denominator*, reduce it, and convert to a decimal.

* Why 12 as a denominator? Because there are 12 inches in a foot.

$$\frac{3}{12} \text{ reduces to } \frac{1}{4}$$

```
      _____
  4 ) 1.00
       8
      ___
      20
      20
      ___
```

3" converts to .25'.

Convert the following inches to their decimal equivalents in feet (to the nearest hundredth).

7. 9" = _____   8. 10" = _____

9. 2" = _____   10. 4" = _____

HOW TO CONVERT A DECIMAL TO A FRACTION

Problem:

You have .25 of a dollar. What fraction of a dollar is that?

Step 1. Say the decimal in words.

.25 is said as "twenty-five hundredths"

Step 2. Write the decimal as a fraction.

.25 is written as a fraction as 25/100

Step 3. Reduce it to its lowest terms.

$$\frac{25}{100} = \frac{25 \div 25}{100 \div 25} = \frac{1}{4}$$

Step 4. Read that .25 converted to a fraction is 1/4. If you have .25 of a dollar, you have 1/4 of a dollar.

SELF-CHECK REVIEW / PRACTICE QUESTIONS 35

Convert the following decimals to their equivalent fractions expressed in lowest terms.

1.   0.5   = _____   2.   0.12 = _____

3.   0.125 = _____   4.   0.30 = _____

5.   0.75 = _____   6.   0.65 = _____

7.   0.80 = _____

Convert the following mixed decimals (whole numbers with decimals) to their equivalent mixed numbers (whole numbers with fractions expressed in lowest terms).

8.   2.8 = _____

9.   9.40 = _____

10. 5.05 = _____

4.0.0   THE METRIC SYSTEM

The metric system is a system of measurement that uses a Base Ten method of determining weight, length, volume and temperature.

4.1.0   INTRODUCTION TO THE METRIC SYSTEM

The ability of the United States to compete in world trade and improve our trade balance is becoming more important and more difficult each day as our competitors get stronger. Since over 95 percent of the world uses the Metric System of measure, the absence of metric in U.S. goods and services can directly reduce our ability to succeed in world markets.

In the 1988 Omnibus Trade and Competitiveness Act, Congress provided us (in part) with the following national policy:

"It is therefore the declared policy of the United States -

(1)   to designate the Metric System of measurement as the preferred

system of weights and measures for United States trade

and commerce.

(2)   to require that each Federal agency, by a date certain and to

the extent economically feasible by the end of fiscal year 1992,

use the Metric System of measurement in its procurements, grants,

and other business-related activities ..."

It is almost certain that the company you are working for is doing business or will be doing business with the U.S. Government. Now is the time to learn the Metric System.

You may already be familiar with some of the common metric units. Have you purchased a 2-liter bottle of soda yet? Run a 10K (Kilometer) lately?

4.1.1 Units Of Weight, Length, Volume, Temperature

The name of each metric measurement *(Figures 21, 22, 23, and 24)* tells you two things:

1. What type of measurement it is (the basic unit):

Grams = Weight

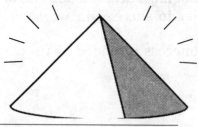

Figure 21. A Gram of Gold

Meters = Length

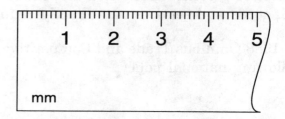

Figure 22. A Meter Stick

Liters = Volume

Figure 23. A 2-Liter Soda

Celsius = Temperature

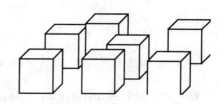

Figure 24. Ice Cubes Freeze at 0° Celsius

2. Its size (in relation to the basic unit, e.g. meter):

deka (da)	=	10	**deci** (d)	=	.1
hecto (h)	=	100	**centi** (c)	=	.01
kilo (k)	=	1,000	**milli** (m)	=	.001
mega (M)	=	1,000,000	**micro** (μ)	=	.000001

The most common prefixes are kilo, milli, and centi. Hecto is used mainly with meters in calculating land size. Mega and micro are used in scientific and engineering measurements.

Here's a way to help you memorize the metric prefixes:

If you won $10, you'd buy a *deck of* cards.   deka = 10

If you won $100, you'd have a *heckuva* good time.   hecto = 100

If you won $1000, you might *keel over*.   kilo = 1,000

If you won $1,000,000, you'd be *mega*-rich.   mega = 1,000,000

Make up your own memorizing tool if you like.

SELF-CHECK REVIEW / PRACTICE QUESTIONS 36

1. A **deka**gram is how many grams?   _____

2. A **hecto**gram is how many grams?   _____

3. A **kilo**gram is how many grams?   _____

4. A **mega**gram is how many grams?   _____

And now a memory tool for the smaller units:

> *Desi sent Milli to Micronesia.*

From small to smaller, that's

Desi	deci =	.1	(1-tenth)
sent	centi =	.01	(1-hundredth)
Milli	milli =	.001	(1-thousandth)
Micronesia	micro =	.000001	(1-millionth)

SELF-CHECK REVIEW / PRACTICE QUESTIONS 37

1. A **deci**gram is how much of a gram?   _____

2. A **micro**gram is how much of a gram?   _____

3. A **milli**gram is how much of a gram?   _____

4. A **centi**gram is how much of a gram?   _____

Match the metric suffix with the type of measurement associated with it. Write the letter in the blank.

5. Gram   _____   A.   Length

6. Liter   _____   B.   Temperature

7. Meter   _____   C.   Weight

8. Celsius   _____   D.   Volume

These charts *(Figures 25, 26, 27, and 28)* will make it easy to see the comparisons:

WEIGHT UNITS

1 Kilogram	=	1000 Grams
1 Hectogram	=	100 Grams
1 Dekagram	=	10 Grams
1 Gram	=	1 Gram
1 Decigram	=	0.1 Gram
1 Centigram	=	0.01 Gram
1 Milligram	=	0.001 Gram

Figure 25. Metric Weight Units

A **Gram** is a little more than the weight of a paper clip.

A **Kilogram** is a little more than two pounds (about 2.2 pounds).

TEMPERATURE UNITS

0° C	Freezing point of water (32° F)
10° C	A warm winter day (50° F)
20° C	A mild spring day (68° F)
30° C	Quite warm - almost hot (86°F)
37° C	Normal body temperature (98.6° F)
40° C	Heat wave conditions (104° F)
100° C	Boiling point of water (212° F)

Figure 26. Metric Temperature Units

VOLUME UNITS

1 Kiloliter	=	1000 Liters
1 Hectoliter	=	100 Liters
1 Dekaliter	=	10 Liters
1 Liter	=	1 Liter
1 Deciliter	=	0.1 Liter
1 Centiliter	=	0.01 Liter
1 Milliliter	=	0.001 Liter

Figure 27. Metric Length Units

Five **Milliliters** make a teaspoon.

A Liter is a little larger than a quart (about 1.06 quarts).

1 Kilometer	=	1000 Meters
1 Hectometer	=	100 Meters
1 Dekameter	=	10 Meters
1 Meter	=	1 Meter
1 Decimeter	=	0.1 Meter
1 Centimeter	=	0.01 Meter
1 Millimeter	=	0.001 Meter

Figure 28. Metric Length Units

A **Millimeter** is about the size of the diameter of a paper clip wire.

A **Centimeter** is a little more than the width of a paper clip (about 0.4 inch).

A **Meter** is a little longer than a yard (about 1.1 yards).

A **Kilometer** is a little over a half of a mile (about 0.6 of a mile).

4.1.2 Introduction To Metric Conversions

Sometimes it may be necessary to change from one unit of measurement to another - say from inches to yards or from centimeters to meters.

In the Standard Measurement System, this is a bit cumbersome. If you want, for example, to change from inches to yards, you must first divide the number of inches by 12 (the number of inches in a foot) and then divide that number by 3 (the number of feet in a yard).

How many yards are in 72 inches? See *Figure 29.*

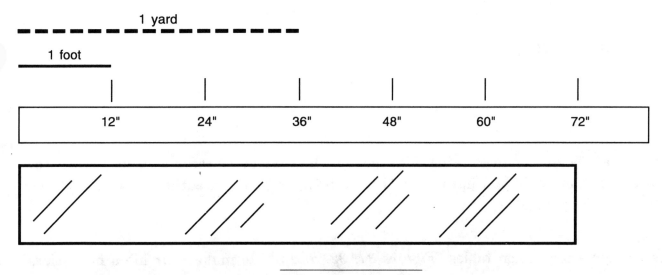

Figure 29. 72 Inches

There are 2 yards in 72 inches.

In the Metric System, you would simply move the decimal point since the system is built on multiples of 10.

How many meters are there in 72 centimeters? See *Figure 30.*

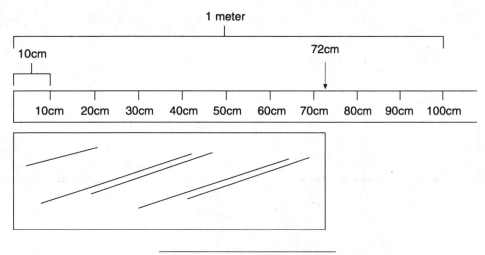

Figure 30. 72 Centimeters

Since 1 centimeter = .01 meter (move decimal 2 places to the left),

72 centimeters = .72 meters.

There are .72 meters in 72 centimeters.

This problem is similar to working with money. If you had 72 cents in your pocket, how much of a dollar would you have? You'd have $.72, or 72 hundredths of a dollar.

Here are some conversion tables *(Figures 31, 32, and 33)* to make your life a bit easier.

HOW TO CONVERT UNITS OF LENGTH

Metric to English			English to Metric		
From	**Multiply By**	**To Obtain**	**From**	**Multiply By**	**To Obtain**
Meters	39.37	Inches	Inches	2.54	Centimeters
Meters	3.2808	Feet	Inches	0.0254	Meters
Meters	1.0936	Yards	Inches	25.4	Millimeters
			Miles	1,609,344	Millimeters
Centimeters	0.3937	Inches	Feet	0.3048	Meters
Millimeters	0.03937	Inches	Feet	30.48	Centimeters
Kilometers	0.6214	Miles	Yards	0.9144	Meters
			Yards	91.44	Centimeters
			Miles	1.6093	Kilometers

Figure 31. Metric Conversion: Length

SELF-CHECK REVIEW / PRACTICE QUESTIONS 38

Convert the following.

1. 2 inches = _____ cm   2. 2 meters = _____ feet

3. 10 kilometers = _____ miles   4. 3 miles = _____ kilometers

5. A 32" counter had to be edged. The counter had how many millimeters to edge? _____ millimeters

6. You drove 14 miles to the lumber yard for materials, then 13 miles to the job site. The total trip was how many kilometers?_____ kilometers

7. If the dimensions of the room are 10 ft by 20 ft, what are the dimensions in meters? _____ meters

8. The diameter of a screw is 5mm. What is the diameter in inches? _____ in.

HOW TO CONVERT UNITS OF WEIGHT

Metric to English			English to Metric		
From	**Multiply By**	**To Obtain**	**From**	**Multiply By**	**To Obtain**
Grams	0.0353	Ounces	Pounds	0.4536	Kilograms
Grams	15.4321	Grains	Pounds	453.6	Grams
Kilograms	2.2046	Pounds	Ounces	28.35	Grams
Kilograms	0.0011	Tons (short)	Grains	0.0648	Grams
Tons (metric)	1.1023	Tons (short)	Tons (short)	0.9072	Tons (metric)

Figure 32. Metric Conversion: Weight

SELF-CHECK REVIEW / PRACTICE QUESTIONS 39

Convert the following.

1. 2 grams = _____ounces

2. 2 pounds = _____grams

3. 10 ounces = _____ grams

4. 3 kilograms =_____pounds

5. A stack of tiles weighs 100 lbs. The same stack weighs _____ kilograms?

6. You received a shipment of nails that weighs 25 kilograms. The same shipment weighs _____ pounds.

7. If you weigh 175 pounds, how many kilograms is that? _____kilograms

8. If you weigh 130 pounds, how many kilograms is that? _____ kilograms

9. How many kilograms do *you* weigh? _____ kilograms

HOW TO CONVERT UNITS OF VOLUME

Metric to English			English to Metric		
From	**Multiply By**	**To Obtain**	**From**	**Multiply By**	**To Obtain**
Liters	1.0567	Quarts	Quarts	0.946	Liters
Liters	2.1134	Pints	Pints	0.473	Liters
Liters	0.2642	Gallons	Gallons	3.785	Liters

Figure 33. Metric Conversion: Volume

SELF-CHECK REVIEW / PRACTICE QUESTIONS 40

Convert the following.

1. 2 quarts = _____ liters

2. 2 gallons = _____ liters

3. 10 liters = _____ gallons

4. 2 liters = _____ quarts

5. If there is water flowing through a pipe at a speed of 5 gallons a minute, how many liters per minute flow through the pipe? _____

6. If a gas tank holds 70 liters of gas, how many gallons does it hold? _____

7. The parts cleaner can hold 20 liters of cleaning solvent. This is _____ gallons.

8. Cooling oil is shipped in 55-gallon drums. This is equal to how many liters? _____

4.1.3 Using A Metric Ruler

Metric rulers are often used with blueprints in which most measurements are given either in centimeters or millimeters. In metalworking, it is most common for measurements to be taken in millimeters.

The Metric ruler below *(Figure 34)* is divided into centimeters and millimeters.

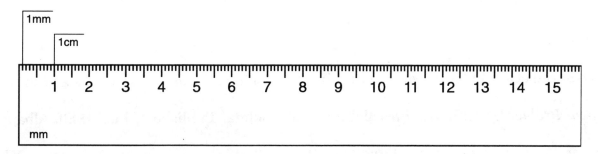

Figure 34. Metric Ruler

SELF-CHECK REVIEW / PRACTICE QUESTIONS 41

Give the measurements for the numbers below in either centimeters or millimeters *(see Figure 35)*, as the problem requires.

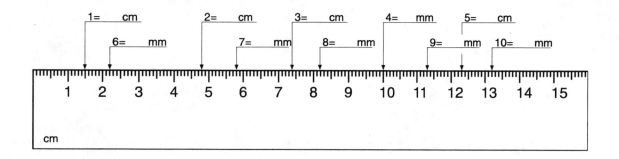

Figure 35. How Many Centimeters or Millimeters?

1. _____ cm   2. _____ cm   3. _____ cm   4. _____ mm

5. _____ cm   6. _____ mm   7. _____ mm   8. _____ mm

9. _____ mm   10. _____ mm

SUMMARY

You have completed the learning and practice portion of this module, *Basic Math*. Your instructor will now provide you with a Module Examination to test your comprehension of this course.

References

For advanced study of topics covered in this Task Module, the following works are suggested:

Practical Problems in Mathematics, John E. Ball, Delmar Publishers, Inc., 1980, Albany, NY

Math for the Industrial Shop, Diane Cheatham, Chatfield College, 1989, Detroit, MI

APPENDIX A: ANSWERS TO SELF-CHECK REVIEW / PRACTICE QUESTIONS

1

1. 3,964
2. tens=1
3. thousands=5
4. units=8
5. hundreds=7
6. ten-thousands=2
7. 85
8. 122
9. 2,497
10. 18,046

2

1. 107
2. 118
3. 136
4. 526
5. 807
6. 1,081
7. 1,611
8. 1,236
9. 13,540
10. 36,891

3

1. 61 people
2. 1,759 bricks
3. 142 tiles
4. 572 screws
5. 37,500 bricks

4

1. 49
2. 9
3. 58
4. 228
5. 109
6. 178

7. 628
8. 709
9. 2,078
10. 1,085 ft.
11. 16 bags
12. 52 workers

5

1. 4 ´ 5; 20
2. product
3. 72
4. 128
5. 46
6. 48
7. 66
8. 84

6

1. 10
2. 15
3. 9
4. 1,808
5. 310
6. 2,584
7. 71,478
8. 87,699
9. 215,871
10. 750 blocks
11. 1,970 watts
12. Electrical: 282 hrs.
13. Masonry: 658 hrs.
14. Pipefitting: 1,128 hrs.
15. Welding: 1,410 hrs.

7

1. 5
2. 9
3. 10 r4
4. 2 r5

5. 5 r4
6. 8 r4
7. 3 r2
8. 5 r5
9. 3 r3
10. 8 r3

8

1. 21 r11
2. 263 r10
3. 302 r2
4. 9 outlets
5. 5 sections
6. 6 rolls
7. 7 hrs.
8. 25 staples
9. 260 tools; $14.00 left
10. 508 sq. ft.

9

1. 69
2. 209
3. 168
4. 1,059
5. 2,356
6. 2,083
7. 1,268
8. 1,769
9. 1,456
10. 980

10

1. 7
2. 9
3. 34
4. 38
5. 37
6. 165

7. 697
8. 850
9. 346
10. 41°
11. Yes; 2"

11

1. 1,808
2. 310
3. 2,584
4. 71,478
5. 87,699
6. 215,871
7. 7,440 screws
8. 286 sq. ft.
9. 216"

12

1. 5
2. 9
3. 6
4. 2 r5
5. 5 r4
6. 8 r4
7. 5 containers
8. 25 assemblies; 2 bolts left over

9. 18 lbs.
10. 49 hrs

13

1. 1"
2. 2-1/2"
3. 1-3/4"
4. 7/8"
5. 3/16"
6. 4-1/4"
7. 1-15/16" or 2"
8. (Check line; should be 3-1/8")
9. (Check line; should be 4-5/16")
10. (Check line; should be 2-1/4")

14

1. 4
2. 4
3. 6
4. 2
5. 2
6. 12
7. 4
8. 48
9. 8
10. 8

15

1. 1/4
2. 1/2
3. 1/4
4. 1/2
5. 3/8
6. 1/8
7. 1/2
8. 3/4
9. 1/16
10. 9/32

16

(Correct answer may be any multiple of the lowest common denominator.)

1. 6 or 18
2. 20
3. 42
4. 72
5. 12 or 24 or 36
6. 63
7. 88
8. 32 or 64 or 96 or 128 or 512
9. 12 or 24

17

1. 3/8
2. 13/18
3. 13/20
4. 17/12 or 1-5/12

5. 2/5
6. 41/24 or 1-17/24
7. 19/12 or 1-7/12
8. 29/24 or 1-5/24
9. 3/4

18

1. 1-3/10
2. 3-3/8
3. 5-9/10
4. 5-2/5
5. 15-1/3
6. 9-1/7
7. 15-1/5
8. 7-2/3
9. 6-3/4

19

1. 1/16
2. 1/9
3. 7/20
4. 7/60
5. 1/24
6. 3/20
7. 3/16
8. 13/24
9. 7/15

20

1. 7-1/4
2. 11-3/8
3. 5-33/40
4. 3/64" thick
5. 1-21/64"
6. 1/64"
7. 7/16"
8. 16-5/6'
9. Min: 7-13/32"   Max: 7-15/32"
10. 57-1/2 cu. yds.

 # 21

1. 1/2
2. 3/8
3. 3/28
4. 21/32
5. 1/8
6. 1/12
7. 10
8. 3-3/4

 # 22

1. 1/8"
2. 1-1/4"
3. 2"
4. 22 yd.
5. 34'
6. 8
7. $4

 # 23

1. 0.9"
2. 1.4"
3. 2.6"
4. 3.3"
5. 4.2"
6. 4.5"

 # 24

1. four tenths
2. five tenths
3. twelve hundredths
4. twenty-five hundredths
5. two hundred forty-five thou-sandths
6. two and five tenths
7. six and twelve hundredths
8. five and twenty-five thousandths
9. 0.004
10. 0.18

 # 25

1. 3, 1 ,2, 4
2. 4, 1, 3, 2
3. 4, 2, 3, 1
4. 3, 2, 1, 4

 # 26

1. 11.0
2. 5.60
3. 7.38
4. 10.48
5. 92.83
6. 0.3284" thick
7. 4.4 tons
8. 1.21 tons

 # 27

1. 0.9"
2. 5.186"
3. 51.48 watts
4. 35.98 watts
5. 87.46 watts
6. No
7. 608 lbs
8. $12.25
9. $26.25
10. $5.70
11. $44.20

 # 28

1. 2.52
2. 0.25
3. 0.02
4. 0.40
5. 5.2
6. 0.02

 # 29

1. 20
2. 216.87
3. 220.0
4. 4,080
5. 520
6. 2,200

 # 30

1. 4.91
2. 10.51
3. 22
4. 40.8
5. 52
6. 2.2

 # 31

1. 25 pieces
2. 15.1
3. 15.3
4. 14.9
5. 15.5
6. Best mileage is car D.
7. $4.06
8. $3.57
9. Best buy is company B.

 # 32

1. 53.74
2. 0.91
3. 629.52
4. 1.98
5. 38.40
6. 6.07
7. 479.21
8. 31.13
9. 78.68
10. 3.78
11. 572.17
12. 2.68

33

1. 62%
2. 62.5%
3. 70%
4. 634%
5. 12%
6. 6,420%
7. 0.72
8. 0.125
9. 2.00
10. 0.34
11. 0.355
12. 3.50

34

1. 0.25
2. 0.75
3. 0.125
4. 0.3125
5. 0.3125
6. 0.1875
7. 0.75'
8. 0.83'
9. 0.17'
10. 0.33'

35

1. 1/2
2. 3/25
3. 1/8
4. 3/10
5. 3/4
6. 13/20
7. 4/5
8. 2-4/5
9. 9-2/5
10. 5-1/20

36

1. 10
2. 100
3. 1,000
4. 1,000,000

37

1. .1
2. .000001
3. .001
4. .01
5. C
6. D
7. A
8. B

38

1. 5.08 cm
2. 6.5616 ft.
3. 6.214 miles
4. 4.8279 km.
5. 812.8 mm
6. 43.4511 km
7. 3.048 m ´ 6.096 m (9.144 m total-optional)
8. 0.19685"

39

1. .0706 oz.
2. 907.2 grams
3. 283.5 grams
4. 6.6138 lbs.
5. 45.36 kg
6. 55.115 lbs.
7. 79.38 kg
8. 58.968 kg
9. (Variety of answers-this is your weight in kilograms)

40

1. 1.892 liters
2. 7.57 liters
3. 2.642 gallons
4. 2.1134 quarts
5. 18.925 liters per minute
6. 18.494 gallons
7. 5.284 gallons
8. 208.175 liters

41

1. 1.5 cm
2. 4.8 cm
3. 7.4 cm
4. 100 mm
5. 12.3 cm
6. 22 mm
7. 58 mm
8. 82 mm
9. 113 mm
10. 132 mm

WHEELS OF LEARNING USER UPDATES

The NCCER makes every effort to keep these manuals up-to-date and free of technical errors. We appreciate your help in this process. If you have an idea for improving this manual, or if you find an error, a typographical mistake, or an inaccuracy in the *Wheels of Learning*, please write us, using this form or a photocopy. Be sure to include the exact module number, page number, a description of the problem, and the correction, if possible. We'll do our best to correct it in later editions. Thank you for your assistance.

Write: *Wheels of Learning*
National Center for Construction Education and Research
P.O. Box 141104
Gainesville, FL 32614-1104
Fax: 352-334-0932

WHEELS OF LEARNING USER UPDATE

Please let us know if you have found an inaccuracy, error, or other problem in a *Wheels of Learning* manual. Use this form or write us a letter. Please be sure to tell us the exact module name and module number, the page number, and the problem. Thanks for your help.

Craft _____   Module Name _____

Module Number _____   Page Number(s) _____

Description of Problem _____

_____

_____

(Optional) Correction of Problem _____

_____

_____

(Optional) Your Name and Address _____

_____

Introduction to Hand Tools

Module 00103

NATIONAL
CENTER FOR
CONSTRUCTION
EDUCATION AND
RESEARCH

INTRODUCTION TO HAND TOOLS

Objectives

Upon completion of this module, the trainee will be able to:

1. Recognize basic hand tools used in the construction trade.
2. Safely use these basic hand tools.
3. Have an awareness of basic maintenance procedures on these hand tools.

Prerequisites

Successful completion of the following Task Modules is required before beginning study of this Task Module: NCCER Task Module 00101, *Basic Safety;* NCCER Task Module 00102, *Basic Math.*

How to Use this Manual

During the course of completing this module, you will be taught and will practice safe use of basic hand tools required for the construction trade. *Self-Check Review / Practice Questions* will follow the introduction of most tools. The answers to the written exercises are found in Appendix A of this manual, titled *Answers to Self-Check Review / Practice Questions.* Your instructor will observe the performance of the *Performance / Laboratory Exercise* to ensure proper tool use.

New terms will be introduced in **bold** print. The definition of these terms can be found in the front of this manual, under *Trade Terms Introduced in this Module.*

Required Student Materials

Equipment

1. Hammer (claw)
2. Screwdriver (straight blade and Phillips head)
3. Sledgehammer
4. Ripping bars and nail pullers
5. Wrench (adjustable and nonadjustable)
6. Pliers and wire cutter
7. Level (spirit level)
8. Square (combination square)
9. File (rasp-cut, curved-tooth, single-cut)
10. Ruler and measuring tape (steel rule, wooden folding rule, steel measuring tape)
11. Saw (crosscut, rip saw)
12. Bench vise
13. C-clamp
14. Chisel and punch
15. Plumb bob
16. Sockets and ratchets
17. Torque wrench
18. Wedge
19. Chalk line
20. Utility knife
21. Chain falls and come along
22. Wire brush

Materials

1. Student Manual (this book)
2. Safety glasses and gloves
3. Sharpened pencil
4. Scrap metal of various thicknesses
5. Scraps of wood; some hard, some soft
6. Wooden stake (at least 4') with tapered end
7. 2 or 3 pieces of wire of varying gauges
8. Scrap linoleum tiles
9. Prepared board with screws
10. 5 or 6 nails
11. Slotted and Phillips-head screws
12. Varying sizes of nuts and bolts, in toolbox

Course Map Information

This course map shows all of the *Wheels of Learning* task modules in the Core Curricula. The suggested training order begins at the bottom and proceeds up. Skill levels increase as a trainee advances on the course map. The training order may be adjusted by the local Training Program Sponsor.

Course Map: Core Curricula, Introduction to Hand Tools

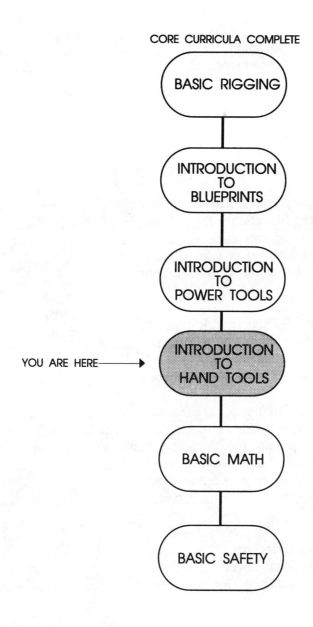

CORE CURRICULA COMPLETE

BASIC RIGGING

INTRODUCTION TO BLUEPRINTS

INTRODUCTION TO POWER TOOLS

YOU ARE HERE ⟶ INTRODUCTION TO HAND TOOLS

BASIC MATH

BASIC SAFETY

TABLE OF CONTENTS

Trade Terms Introduced in This Module:

Alloy: A substance that is a mixture of two or more metals.

Anvil: An iron or steel block on which metal objects are hammered into shape.

Ball peen hammer: A hammer with a flat face used for striking and a rounded front section.

Bell faced hammer: A claw hammer with a slightly rounded, or convex, face.

Beveled: Sloped.

Cast: To form by pouring or pressing into a mold. Weaker than dropped-forged.

Dropped-forged: Pounded, heated metal. Stronger than cast.

Dowel: A (usually) round pin that fits into a corresponding hole to fasten or align two adjacent pieces.

Fastener: A device used to attach or secure materials of various types one to another thing, e.g., bolt, clasp, hook, lock etc.

Insulated: Covered with a nonconductive material to prevent the passage of electricity.

Kerf: Cut or channel made by a saw.

Level: Perfectly horizontal.

Mallet: A short handled hammer, usually with a cylindrical head of wood.

Plumb: Perfectly vertical.

Points: Teeth per inch on a handsaw.

Tempered: A heat treatment used to create or restore hardness in steel.

Weld: A union or joint produced by welding.

1.0.0   INTRODUCTION

Every profession has its tools. A surgeon has the scalpel, a teacher has books, a computer operator has a computer. The better the worker can use and maintain the tools, the better he or she is in their craft.

In the construction trade, there is a collection of tools that are used across the trade. Tools such as hammers, screwdrivers, and pliers are used by almost all craftworkers. Though the use of some of these tools may be obvious, proper maintenance and safety are equally important in tool use.

This module will provide you the opportunity to safely use and maintain 22 of the most commonly used hand tools of the construction trade.

1.1.0   SAFETY

In order to work safely, it is necessary to think safety. A fundamental procedure of using any tool is knowing how it works, what function it has, and some of the possible dangers of using it incorrectly. Before using any tool, read the procedures and any safety precautions. Make sure the tool is in good shape.

WARNING!   Always wear protective eye gear and safety gloves when using tools!

2.0.0   HAMMERS

Hammers are made in various sizes and weights, each developed for specific types of work. Two of the most common types of hammers are the **claw hammer** and the **ball peen hammer**.

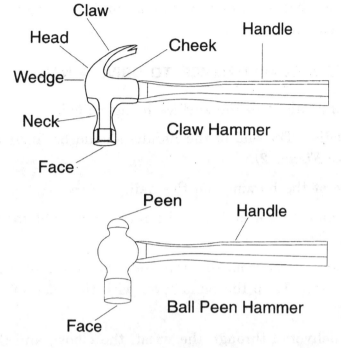

Figure 1.   The Claw Hammer and Ball Peen Hammer

2.1.0   DESCRIPTION AND SELECTION

The claw hammer *(Figure 1)* has a steel head and a wooden handle. It is used for driving nails, wedges, and **dowels**. The claw is used to pull nails out of wood. Other parts of the hammer are pictured above.

The face of the hammer may be flat or rounded. The flat, or plain face, claw hammer is easier to learn to drive nails with. But keep in mind that with this type of hammer it is more difficult to drive the head of the nail flush, or even, with the surface of the work without leaving hammer marks.

A claw hammer with a slightly rounded, or convex, face is called **bell faced**. When used by an expert, it can drive the nailhead flush without damaging the surface of the work.

The ball peen hammer *(Figure 1)* has a flat face used for striking and a rounded section used to align brackets, and to drive out bolts. It is also used with chisels and punches (discussed later in this course). The ball peen is classified by weight and weighs from 6 ounces to 2-1/2 pounds.

The quality of hammers is an important factor to consider. Hammers made from tough **alloy** and **dropped-forged** steel are the best hammers. Inexpensive hammers are usually made with **cast** heads and are very brittle. They are not suited for construction work because they tend to chip and break. It is in your best interest *not* to use a hammer with a cast head. A chip could easily break off and become a dangerous object. It could cause serious injury to you or your partner.

2.2.0   HOW TO USE A CLAW HAMMER TO DRIVE A NAIL

Step 1   Hold the nail straight to the surface being nailed.

Step 2   Grip the handle. The end of the handle should be flush with the lower edge of the palm *(See Figure 2)*.

Step 3   Rest the face of the hammer on the nail.

Step 4   Draw the hammer back and give the nail a few light taps to start the nail and to determine the aim.

Step 5   Take the fingers away from the nail and drive the nail firmly with the center of the hammer face. Keep the hand level with the head of the nail and strike the face squarely.

The blow is delivered through the wrist, the elbow, and the shoulder.

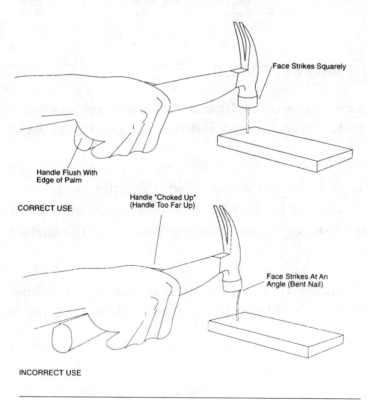

Figure 2.   Correct and Incorrect Use of the Hammer

2.3.0 HOW TO USE A CLAW HAMMER TO PULL A NAIL

Step 1   Slip the claw of the hammer under the nail head and pull until the handle is nearly vertical (straight up) and the nail partly drawn *(See Figure 3)*.

Step 2   Pull the nail out vertically from the wood.

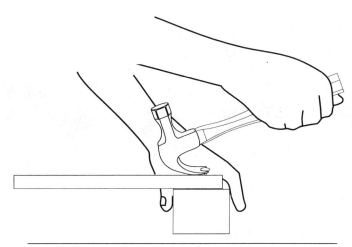

Figure 3.   Using a Claw Hammer to Pull a Nail

2.4.0 HOW TO USE A BALL PEEN HAMMER

Step 1   Grip the handle. The end of the handle should be flush with the lower edge of the palm with the hammer face parallel with the work.

Step 2   Use the face for hammering and the ball for rounding off rivets and other similar jobs.

2.5.0 SAFETY AND MAINTENANCE CONSIDERATIONS

1.   KEEP FOCUSED on your work.
2.   Be sure that the handle of the hammer is without splinters.
3.   Be sure the handle is set correctly and securely in the head of the hammer.
4.   Be sure to replace all cracked and broken handles.
5.   Be sure that the face of the hammer is clean.
6.   Blurred, chipped or mushroomed heads should be discarded.
7.   Be sure to hold the handle properly. Grasp the handle firmly near the end and strike the nail squarely.
8.   Be sure not to strike the cheek of the hammer.
9.   Do not use a hammer with a cast head. A chip could easily break off and become a dangerous object, causing serious injury to you or your partner.

SELF-CHECK REVIEW / PRACTICE QUESTIONS 1

1. The most commonly used hammer is the _____C_____ hammer.

   A. Beak   B. Bell   C. Claw   D. Wedge

2. The claw of the hammer is used to: _____A_____

   A. Pull nails out of wood   B. Scrape paint from walls
   C. Remove loose wires   D. Drive large metal spikes

3. The bell-faced (rounded) claw hammer can drive the nailhead flush without damaging the surface of the work. _____B_____

   A. True   B. False

4. Which is the safer type of hammer? _____

   A. Tough alloy and dropped-forged
   B. Cast head

5. When driving a nail, you should grip the handle of the hammer close to its end. _A_

   A. True   B. False

PERFORMANCE / LABORATORY EXERCISE

Supplies needed:
   Claw hammer,
   5 or 6 nails,
   scrap piece of wood.

Try this:
   1. Hammer 5 or 6 nails into the board.
   2. Remove 3 of the nails.

Check to be sure that:

1. You hold the handle properly. Grasp the handle firmly near the end and strike the nail squarely.
2. The nails go into the wood straight; that is, that you do not bend them in the process of hammering.
3. The nail heads are flush against the wood once driven.
4. The nail is pulled out vertically to the wood.

3.0.0 SCREWDRIVER

A screwdriver is used to tighten and remove screws. It is identified by the type of screw it fits. *Figure 4* shows two common types of screw heads.

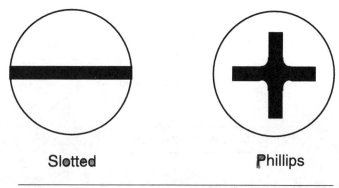

Slotted   Phillips

Figure 4. Two Common Types of Screw Heads

The most common screwdrivers are the **straight–blade** and the **Phillips–head** screwdrivers. These are the two we will cover in this module *(See Figure 5)*.

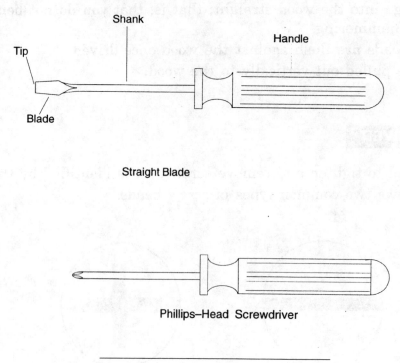

Figure 5. Basic Screwdriver Types

3.1.0   DESCRIPTION AND SELECTION

The screwdriver handle is designed to provide a firm grip. The screwdriver shank is the hardened metal portion between the handle and blade. The shank is made to withstand a large amount of twisting force. These shanks may be round, square, or some other shape.

The blade of a screwdriver is the formed end that fits into a screw head. Industrial screwdriver blades are made of **tempered** steel to resist wear and to prevent bending and breaking.

It is important to select the proper type of blade for the screw head being used. It should fit snugly into the screw head and not be too long, short, loose, or tight. You run the risk of damaging the screwdriver or the screw head with incorrect sizing *(See Figure 6)*.

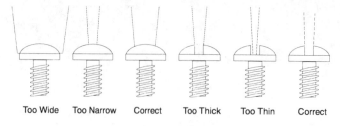

Figure 6. Always Use the Right Type of Screwdriver Blade for the Screw Head

3.2.0 HOW TO USE A SCREWDRIVER

Step 1   Select the proper type of blade for the screw head (*See Figure 5).*

Step 2   Make sure the screwdriver fits the screw correctly, as shown in *Figure 6.*

Step 3   Position the shank vertically (straight up) to your work.

Step 4   Apply firm, steady pressure to the screw head and turn:

- Clockwise to tighten (hint: RIGHT is TIGHT!)
- Counterclockwise to loosen (hint: LEFT is LOOSE!)

CAUTION   When starting the screw, it is easy to hurt your fingers if the blade slips. Work with caution.

3.3.0 SAFETY AND MAINTENANCE CONSIDERATIONS

CAUTION   Never use the screwdriver as an electrical tester!

1. Never put any part of your body (or anyone else's!) in front of the blade.
2. Always point the screwdriver blade away from you when working close to your body.
3. Keep the screwdriver free of dirt, grease, and grit so the blade will not slip from the screw head slot.
4. Never use the screwdriver as a punch, chisel, or pry bar.
5. Do not use a screwdriver near live wires.
6. Do not expose a screwdriver to excessive heat.
7. File a worn tip to restore a straight edge.
8. Do not use a screwdriver that has a worn or broken handle.

SELF – CHECK REVIEW / PRACTICE QUESTIONS 2

1. Two of the more common type of screwdrivers are the:___*D*___

    A. Phillips-Head and Sharp Point   B. Phylladendra and Straight Edge
    C. Phillips and Blade Runner   D. Phillips Head and Straight Blade

2. A screwdriver handle is designed to:___D___

    A. Hammer nails   B. Resist grease
    C. Pry open a jammed door   D. Provide a firm grip

3. Which of these three *(Figure 7)* are correctly fitted?

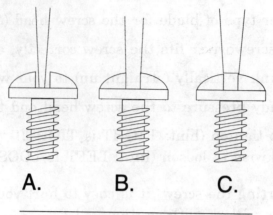

A.   B.   C.

Figure 7. Which is the Correct Fit?

4. It is OK to use the screwdriver as a chisel or pry bar.

    A. True   B. False

PERFORMANCE / LABORATORY EXERCISE

Supplies needed:
    Straight-blade screwdriver
    2 or 3 straight-blade screws
    Phillips-head screwdriver
    2 or 3 Phillips-head screws
    1 piece of "soft" wood (at least 6" long, 1/2" thick)

Try this:
1. Use the straight blade screwdriver to drive in 2 or 3 straight-blade screws.
2. Use the Phillips-head screwdriver to drive in 2 or 3 Phillips head screws.
3. Remove (unscrew) 1 of each type.

Check to be sure that:
1. You are using the right type of screwdriver blade for the screw head.
2. The screwdriver fits the screw correctly.
3. You position the shank vertically (straight up) to your work.

4.0.0 SLEDGEHAMMERS

The sledgehammer is a heavy-duty tool used for driving posts or other large stakes. The head of the sledgehammer is made of high carbon steel and weighs from 2 to 20 pounds. The shape of the head depends on the job the sledgehammer was designed to do.

4.1.0 DESCRIPTION AND SELECTION

There are many types of sledgehammers. Pictured in *Figure 8* are two types: the Double Face and the Cross Peen.

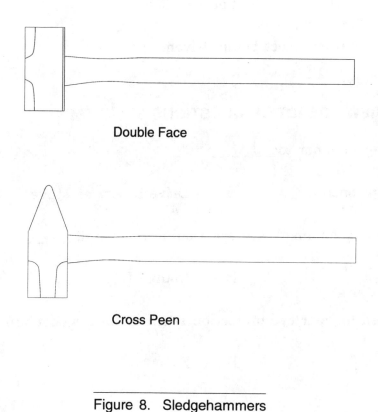

Double Face

Cross Peen

Figure 8.   Sledgehammers

4.2.0 HOW TO USE A SLEDGEHAMMER

Step 1   WEAR EYE PROTECTION. Safety gloves are also recommended.

Step 2   Hold the sledgehammer with both hands.

Step 3   Be sure you have enough room to swing. Check behind you.

Step 4   Position yourself so that the object to be driven is directly in front of you.

Step 5   Deliver striking blows from over your head.

4.3.0 SAFETY AND MAINTENANCE CONSIDERATIONS

1. ALWAYS WEAR EYE PROTECTION WHEN USING A SLEDGEHAMMER. Safety gloves are recommended.
2. Always check behind you to make sure you have enough room to swing.
3. Be sure that the handle is secured properly at the head.
4. Be sure to replace cracked or broken handles.
5. Use the proper force for the job.
6. Keep hands away from object being driven.

SELF-CHECK REVIEW / PRACTICE QUESTIONS 3

1. You use a sledgehammer to:___A-B___

   A.   Break up concrete   B.   Drive a post or stake   C.   Hammer nails

2. Two types of sledgehammers are the Double Face and:___B&C___

   A.   Ball Point   B.   Double Edged   C.   Cross peen

3. You do not need to wear eye protection when using a sledgehammer.___B___

   A.   True   B.   False

A variety of tools are made for ripping and prying apart woodwork and pulling nails. This section discusses **ripping bars** and **nail pullers.** *(See Figure 9.)*

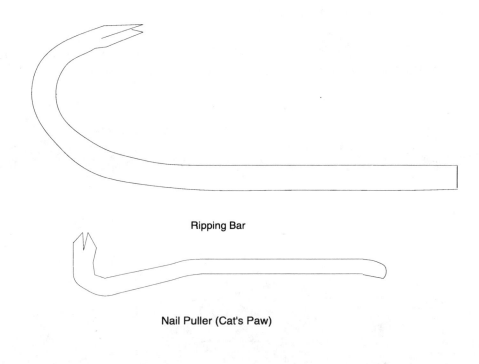

Ripping Bar

Nail Puller (Cat's Paw)

Figure 9. Ripping Bar and Nail Puller

5.1.0   DESCRIPTION AND SELECTION

A type of ripping tool is the **ripping bar**, or pinch bar. The ripping bar is 24 to 36 inches long, with a deeply curved nail claw at one end and an angled, wedge-shaped face at the other. This bar is used for heavy duty dismantling of woodwork, as in tearing apart building frames or concrete forms.

Nail–pulling tools generally are of two types: **"cat's paw"** nail bars and **chisel bars.** The cat's paw is a straight steel rod with a curved claw at one end. It is used most often for pulling nails that have been driven flush with, or slightly below, the surface of the wood. A chisel bar has two claws, one at either end, and is ground to a chisel-like **bevel** on both ends. This bar can be used to pull nails, much in the same fashion that a claw hammer is used, and can also be driven into wood to split and rip apart the pieces.

5.2.0   HOW TO USE A RIPPING BAR

1. Use the angled prying end to force wood members apart.
2. Use the heavy claw to pull large nails and spikes.

5.3.0   HOW TO USE A NAIL PULLER (CAT'S PAW)

1. Drive the claw into the wood, astride the nail head (*See Figure 10*).
2. Pull the handle of the bar to lift the nail out of the wood.

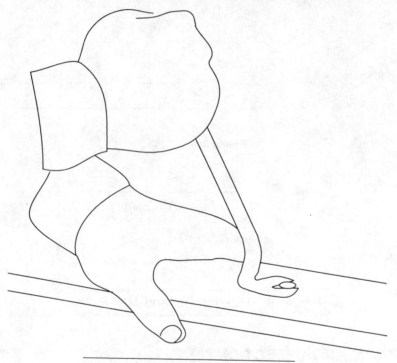

Figure 10. Using the Nail Puller

5.4.0   SAFETY AND MAINTENANCE CONSIDERATIONS

1. For nail pulling: be sure the material housing the nail is braced securely before pulling the nail, to prevent it from hitting you in the face.
2. For ripping: use two hands so as to keep even pressure on your back as you pull.

1.   The ripping bar is used for:___*B*___

    A. Gripping large metal objects for demolition
    B. Heavy dismantling of woodwork
    C. Hammering nails

2.   Two types of nail pulling tools are the:___*B*___

    A. The dog's paw and the chisel bar
    B. The cat's paw and the chisel bar

3.   A nail puller is used to:___*C*___

    A. Drive nails
    B. Break up concrete
    C. Pull nails and rip wood

6.0.0   WRENCHES

Wrenches are used to hold and turn screws, nuts, bolts, and pipe that have hexagonal (**six-sided**) heads. There are many types of wrenches, but they can be divided into two general categories: **nonadjustable** and **adjustable**.

6.1.0   DESCRIPTION AND SELECTION

Nonadjustable wrenches, *Figure 11*, include the open-end wrench, the box-end wrench, and the combination wrench.

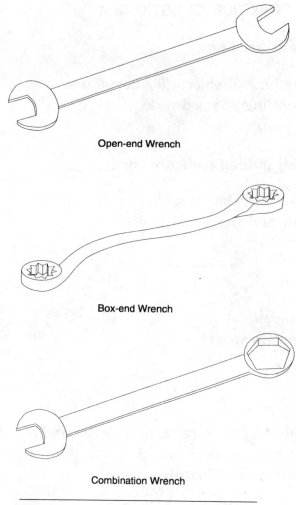

Open-end Wrench

Box-end Wrench

Combination Wrench

Figure 11. Nonadjustable Wrenches

The **open-end** wrench is one of the simplest to use. It has an opening at each end that designates the size of the wrench. Very often, two different sizes are combined in one wrench, such as 7/16" and 1/2".

These dimensions measure the distance between the flats of the wrench and the distance across the head of the fastener used.

Box-end wrenches form a continuous circle around the head of a fastener. The ends come in 6-point or 12-point. The handles come in varying lengths. Common sizes range from 3/8" to 1-5/6". This wrench is safer to use than an open-end wrench because it will not slip off the sides of certain kinds of bolts.

Combination wrenches are, as the name implies, a combination of the two other wrench types. One of the ends of the combination wrench is open and the other is closed, or box-end. Combination wrenches can be used to speed work, as you don't have to change wrenches for two types of jobs! The lengths of the handles on these wrenches vary.

Adjustable wrenches are also used to tighten nuts and bolts *(See Figure 12)*. They have one fixed jaw and one movable jaw. The adjusting nut on the head of the wrench joins the teeth in the wrench body and moves the adjustable jaw. These wrenches come in lengths from 4" to 8" and open from 1/2" to 2-1/16".

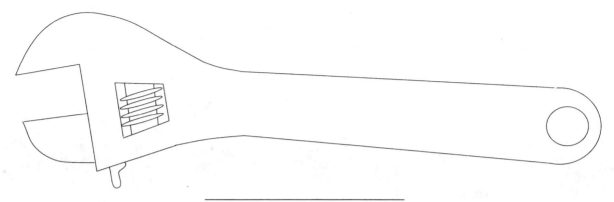

Figure 12. Adjustable Wrench

6.2.0   HOW TO USE NONADJUSTABLE WRENCHES

Step 1   Always use correct size wrench for the nut or bolt.

Step 2   Pull the wrench toward you. Pushing the wrench can cause injury.

6.3.0   HOW TO USE ADJUSTABLE WRENCHES

Step 1   The adjustable wrench should only be used if open-end or box-end wrenches are not available or if the part to be turned is an unusual size.

Step 2   Always use correct size for the nut or bolt.

Step 3   Be sure that the jaws are fully tightened on the work. The wrench should be turned so the heaviest load is always on the stationary jaw *(See Figure 13)*.

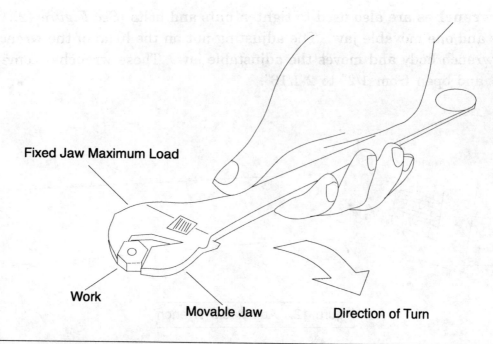

Fixed Jaw Maximum Load

Work

Movable Jaw

Direction of Turn

Figure 13. The Wrench Should be Turned So the Heaviest Load is Always on the Fixed Jaw

CAUTION   If the jaws are improperly adjusted, injury could occur.

Step 4   Check to be sure you have clearance for your fingers.

Step 5   Generally, pull the wrench toward you. Pushing the wrench can cause injury. If you must push on the wrench, use an open hand to avoid a pinch point.

6.4.0   SAFETY AND MAINTENANCE CONSIDERATIONS

1. Keep your attention focused on your work.
2. Pull the wrench toward you. Pushing the wrench can cause injury.
3. Never use the wrench as a hammer.
4. Do not use any wrench beyond its capacity. No extension should be added to increase its leverage. This could cause serious injury.
5. The safest wrench to use is a box-end wrench.
6. Keep adjustable wrenches clean. Do not allow mud or grease to clog the adjusting screw and slide. Oil these parts frequently.
7. Never hammer on a wrench unless it is made for this use, such as a slugging wrench.

SELF-CHECK REVIEW / PRACTICE QUESTIONS 5

1.   Fill in the correct wrench number for each letter. See *Figure 14*.

    A.   Open-end wrench_____

    B.   Box-end wrench_____

    C.   Combination wrench_____

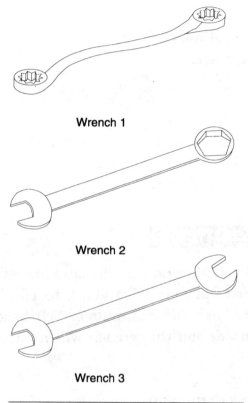

Wrench 1

Wrench 2

Wrench 3

Figure 14. Which Wrench is Which?

2.   The wrenches illustrated in problem #1 above are:_____

    A.   Adjustable wrenches   B.   Nonadjustable wrenches

3.   Always pull the wrench toward you._____

    A.   True   B.   False

4.   The safest wrench to use is the_____wrench.

    A.   Open-end   B.   Box-end   C.   Adjustable

PERFORMANCE / LABORATORY EXERCISE

Supplies needed:

    Combination wrenches (varying sizes)
    Adjustable wrenches (varying sizes)
    Instructor-prepared board with varying size bolts

Try this:

1. Remove 3 varying sized nuts or bolts
2. Practice reinstalling them.

Check to be sure that:

1. You use the correct size wrench for the nut or bolt.
2. The jaws are fully tightened on the work.
3. Always pull the wrench TOWARD you.

7.0.0   PLIERS AND WIRE CUTTERS

Pliers are a special type of adjustable wrench. The jaws are adjustable because the two legs or handles move on a pivot. You will generally use this tool for holding, cutting, and bending wire and soft metals. DO NOT use pliers on nuts or bolt heads as they will round off the edges of the hex (six-sided) head, and the wrench will no longer fit properly.

7.1.0   DESCRIPTION AND SELECTION

Pliers are scissor-shaped tools with jaws. The jaws usually have teeth to help grip objects. Higher quality pliers are made of hardened steel.

Pliers come in many different head styles, depending on their use. The pliers most commonly used are:

- Slip joint (or combination) pliers
- Long nose (or needle nose) pliers
- Lineman (or side cutters) pliers

(See Figure 15)

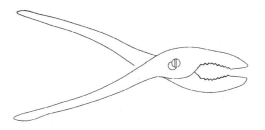

Slip-Joint (Combination) Pliers

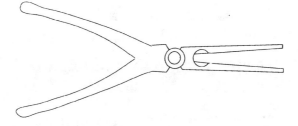

Long-Nose (Needle Nose) Pliers

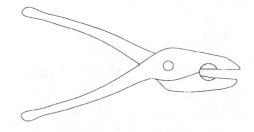

Lineman (Side Cutters) Pliers

Figure 15. Types of Pliers

Slip-joint, or combination pliers are probably the most popular type of plier. They are used for holding and bending wire and gripping and holding objects during assembly operations. The adjustable jaws make it easy to change the distance between the two jaws. There are two jaw settings: one for small material and one for larger materials. Remember that pliers are not wrenches. Do not turn nuts or bolts with them, as you may strip the sides.

The **long-nose**, or needle nose, pliers are great for getting into tight places where other pliers won't reach or for retrieving and holding parts that are too small to be held with the fingers. They are extremely useful for bending angles in wire or narrow metal strips. There is a sharp wire cutter near the pivot.

The **lineman**, or side cutter, pliers have wider jaws than slip-joint pliers. The lineman's primary use is cutting heavy or large-gauge wire and for holding work. The wedged jaws reduce the potential for wire slippage and the hook bend in one handle enhances the non–slip grip.

7.2.0 HOW TO USE SLIP-JOINT (COMBINATION) PLIERS

Step 1   Place the jaws on the object to be held.

Step 2   Squeeze the handles until contact is made.

7.3.0 HOW TO USE LONG-NOSE (NEEDLE NOSE) PLIERS

Step 1   Wear safety glasses.

Step 2   If the pliers do not have a spring opening, hold them with the third finger or little finger positioned inside the handle to open the pliers.

Step 3   Use the sharp cutter near the pivot for cutting wire.

7.4.0 HOW TO USE LINEMAN (SIDE CUTTERS) PLIERS

Step 1   Wear safety glasses.

Step 2   When cutting wire, always point the loose end of the wire down.

Step 3   Cut at a right angle to the wire.

CAUTION   Never rock pliers side to side when cutting. The object being cut could fly in your face.

7.5.0 SAFETY AND MAINTENANCE CONSIDERATIONS

1. Oil pliers regularly to prevent rust and to keep them working smoothly.
2. Pliers are not wrenches. Do not turn nuts or bolts with them.
3. Never expose pliers to excessive heat.
4. Always cut at right angles. Never rock the pliers from side to side or bend the wire back and forth against the cutting blades. Loose wire could fly and injure.
5. Never use pliers as hammers.
6. Never extend the length of the handles for greater leverage. Use a larger pair of pliers instead.
7. Wear safety glasses when cutting wire.
8. Holding pliers close to the end prevents you from pinching your fingers in the hinge.
9. Hold short ends of wires when cutting to avoid flying metal bits.
10. Though the handles may be plastic-coated, they are NOT insulated against electrical shock. DO NOT use around live wires.

SELF-CHECK REVIEW / PRACTICE QUESTIONS 6

1. The reason why pliers should not be used on nuts or bolts is because:_____

   A. They're used only for cutting wire.
   B. They can round off the edges of the hex head.

2. Holding pliers close to the end prevents you from pinching your fingers in the hinge

   A. True_____   B. False_____

3. When cutting wire with pliers, you should rock the pliers side to side for a faster cut.

   A. True_____   B. False_____

4. Most pliers have toothed jaws.

   A. True_____   B. False_____

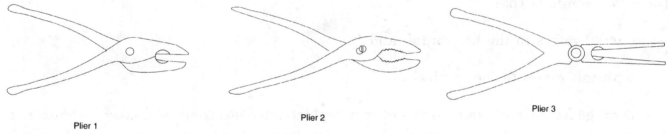

Plier 1

Plier 2

Plier 3

Figure 16.   Which Plier is Which?

5. Fill in the correct plier number beside each letter *(See Figure 16)*.

   _____A. Slip-joint plier
   _____B. Long-nose pliers
   _____C. Side cutters

PERFORMANCE / LABORATORY EXERCISE

Supplies needed:
   Lineman pliers
   2 or 3 pieces of wire of varying gauges
   Safety glasses

Try this:
   Follow the procedures listed above to cut the varying gauged wires in 3" lengths.

Check to be sure that:
   1. The loose end of the wire is pointing down.
   2. You cut at a right angle to the wire.
   3. You hold pliers close to the end.

8.0.0   LEVELS

Levels are used to determine the exactness of both **level** and **plumb**. The difference between these two words is that:

   level refers to the horizontal, and

   plumb refers to the vertical.

Levels range from simple spirit levels to very sophisticated electronic and laser instruments. This module address the more commonly used leveling tool in the construction trade, which is the spirit level.

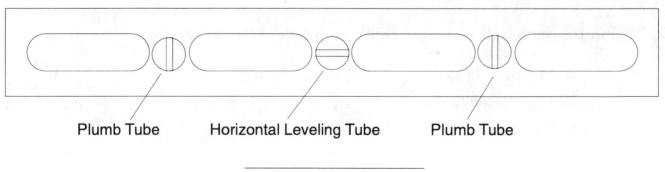

Figure 17. The Spirit Level

8.1.0   DESCRIPTION AND SELECTION

Most levels are made of tough, lightweight metals such as magnesium or aluminum.

The **spirit level** is used to determine if a surface is level or plumb. It has three vials (*see Figure 17*):

- **two end vials** measure plumb (vertical)
- **vial in the center** measures level (horizontal)

The amount of liquid (alcohol) each vial contains is not enough to fill the vial. This creates a bubble when the vial is held in the horizontal or vertical position. When the air bubble centers between the lines, the level is either level (if horizontally placed) or plumb (if vertically placed). (*See Figure 18.*) Spirit levels come in a variety of sizes. The 24" spirit level is considered standard, but there are 28", 48", 72", and 78" spirit levels. The longer the level, the greater the accuracy.

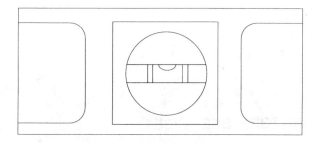

Figure 18.   An Air Bubble Centered Between the Lines Shows Level and/or Plumb

8.2.0   HOW TO USE A SPIRIT LEVEL

Step 1   Place the spirit level on the object to be checked.

Step 2   Observe the bubble. The air bubble should be centered between the lines to achieve level (*see Figure 18*).

- The two end vials measure plumb (vertical).
- The vial in the center measures level (horizontal).

8.3.0   SAFETY AND MAINTENANCE CONSIDERATIONS

1.   Levels are precision instruments. They must be handled with care; do not drop or bump them against other materials.
2.   Keep levels clean and dry to retain their true measure.

1. Plumb refers to_____and level refers to_____. (Fill in the blanks with either vertical or horizontal.)

2. The instrument that has three vials and is used to find if a surface is level or plumb is called a:

   A.   Plumb bob   B.   Spirit level   C.   Line level

3. The two end vials in a spirit level measure:

   A.   Plumb   B.   Level

4. How can you tell that the object being measured is level using the spirit level?

   A.   The liquid turns green
   B.   The air bubble disappears
   C.   The air bubble centers between the lines

PERFORMANCE / LABORATORY EXERCISE

Supplies needed:
    Spirit level

Try this:

   1.   Use the spirit level to check the vertical (plumb) and horizontal (level) accuracy of 3 or 4 objects in the room.

Check to be sure that the air bubble is centered between the lines.

9.0.0   SQUARES

Squares are used for marking, checking, and measuring. (*See Figure 19.*) The type of square to be used depends on the job and the worker's preference. Two common squares are the:

   •   Framing square
   •   Combination square

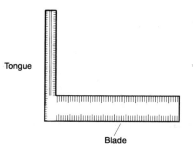

Tongue

Blade

The Framing Square

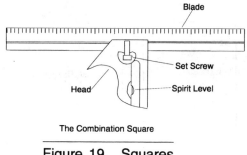

Blade

Set Screw

Spirit Level

Head

The Combination Square

Figure 19.   Squares

9.1.0   DESCRIPTION AND SELECTION

The **framing square** is used mainly for squaring up large pieces. It has a 24" blade and a 16" tongue, forming right angles. The sides of this square are divided into inches and fractions of an inch. Both tongue and blade may be used as a rule and as a straightedge.

The **combination square** has a 12" blade that moves through a head. The head contains a 45- and 90- degree angle measure. Some squares also contain a small spirit level and a carbide scriber for marking metal. The combination square is one of the most useful tools for layout work. With it you can:

*   test work for squareness
*   mark 90- and 45-degree angles
*   check level and plumb surfaces
*   measure lengths and widths
*   use as a straightedge and marking tool

Quality combination squares have all metal parts, a blade that slides freely but which can be clamped securely in position, and a glass tube spirit level that is truly level and tightly fastened.

9.2.0 HOW TO USE A FRAMING SQUARE

To mark a line for cutting:

Step 1   Find and mark the location for the line to be drawn.

Step 2   Place the square so that it lines up with the bottom of the object to be marked (*See Figure 20*).

Step 3   Mark the line and cut off excess material.

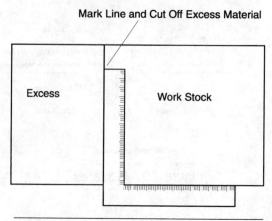

Figure 20.   Marking a Line for Cutting

To check the inside or outside squareness of material (*Figure 21*):

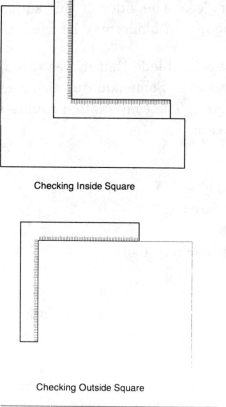

Figure 21.   Checking Squareness

To check flatness of material *(Figure 22)*:

Step 1   Place the edge of the blade of the square on the surface to be checked.

Step 2   Check to see if there is light between the square and the surface of the material being checked.

   If there is, the surface is not flat.

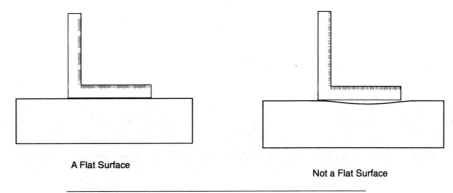

A Flat Surface

Not a Flat Surface

Figure 22. Checking the Flatness of Material

9.3.0   HOW TO USE A COMBINATION SQUARE

To mark 90-degree angles on materials *(Figure 23)*:

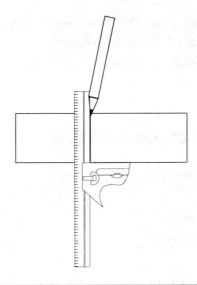

Figure 23. Using a Combination Square to Mark 90 Degree Angle

To mark a 45-degree angle *(Figure 24)*:

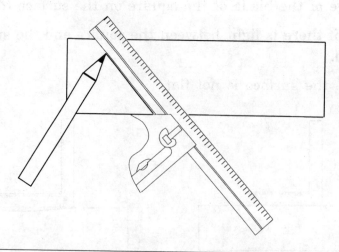

Figure 24. Using a Combination Square to Mark 45-Degree Angle

Note   For marking angles other than 45- and 90- degrees, slide the protractor onto the combination square's blade and dial in the desired angle.

9.4.0   SAFETY AND MAINTENANCE CONSIDERATIONS

1. Be careful not to drop or strike the square hard enough to change the angle between the blade and the head or tongue.
2. Squares must be kept dry to prevent them from rusting.
3. A combination square needs little maintenance. Use a light coat of oil on the blade, and occasionally clean the blade's grooves and the set screw.

SELF-CHECK REVIEW / PRACTICE QUESTIONS 8

1. What would you use a square for?_____
   A. Cutting wire
   B. Calculating areas of a room
   C. Marking, checking, measuring

2. Which is the one thing a combination square does NOT do?_____

   A. Test work for squareness
   B. Mark 90- and 45-degree angles
   C. Tighten screws
   D. Check for level and plumb surfaces
   E. Measure lengths and widths
   F. Function as a straightedge and marking tool

3. You can prolong the life of a square by:_____

    A. Storing it in an outdoor shed
    B. Periodically coating it with oil

PERFORMANCE / LABORATORY EXERCISE

Supplies needed:

> Combination square
> Scrap wood
> Sharpened pencil

Try this:

1. Using the scrap wood, mark a piece of wood for cutting that is 5" wide, 3" long, and is at a 90-degree angle.
2. Using the scrap wood, mark a piece of wood for cutting that is 3-1/2" wide, 5" long, and is at a 45-degree angle.

Check to be sure that:

> You place the square so that it lines up with the bottom of the object to be marked.

10.0.0   RULERS AND MEASURING TAPES

There are three basic types of rulers, or rules, used by the craftworker. These are:

- Steel rule
- Measuring tape
- Wooden folding rule

10.1.0 DESCRIPTION AND SELECTION

When selecting a measuring tool, keep in mind the following:

- It must be accurate
- It should be easy to use
- It should be durable
- The numbers should be easy to read (black on yellow or off-white are good)

The **steel rule** *(Figure 25)* is the simplest and most common of all measuring tools. The flat steel rule is usually 6" or 12" long, but longer sizes are available.

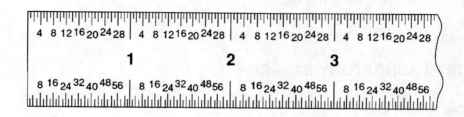

Figure 25. The Steel Rule

Steel rules may be either flexible or non-flexible. The thinner the rule, the more accurately it measures, because the division marks are closer to the work.

Generally, a steel rule has four sets of marks, two on each side of the rule. The longest lines are the inch marks. On one edge, each inch is divided into eight equal spaces of 1/8" each. The other edge is divided into 1/16" spaces. The 1/4" and the 1/2" marks are normally made longer than the smaller division marks to make counting easier.

The opposite side of the steel rule is divided into 32 and 64 spaces to the inch. Each fourth division in the inch is usually numbered for easier reading.

Measuring tapes are available in varying lengths. The shorter tapes are usually made with a curved cross section, so they are flexible enough to roll up but remain rigid when extended. Long, flat tapes should be laid across a surface to prevent sag in the tape.

Steel measuring tapes *(Figure 26)* are usually wound into metal cases. A hook is provided at one end of the tape to hook over the object being measured. They are used to measure longer distances. Better quality tapes have a polyester film bonded to the steel blade to guard against wear. The tape may have a lock that holds the blade in the open position and a rewind spring that returns the blade to the case. Look for ease of reading, also. Black numerals on yellow or off-white backgrounds are easiest to read.

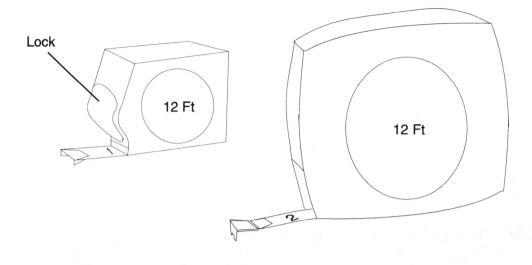

Figure 26.   Steel Measuring Tapes

A **wooden folding rule** *(Figure 27)* is commonly used on the job. This rule is usually marked in sixteenths of an inch on both edges of each side. They are made in 6-foot and 8-foot lengths. Because of its stiffness and its ability to construct angles, a folding rule is preferable to a cloth tape when marking off vertical distance.

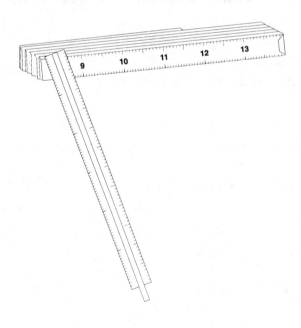

Figure 27.   Wooden Folding Rule

10.2.0 HOW TO USE A STEEL TAPE

Step 1   Pull the tab out to the desired length.

Step 2   Place the hook over the edge of the material being measured. Lock the tape if necessary (use lock button on holder, usually located above tape).

Step 3   Mark or record the measurement.

Step 4   Unhook the tape from the edge.

Step 5   Rewind the tape by pressing the rewind button (usually located on side of tape holder).

10.3.0 SAFETY AND MAINTENANCE CONSIDERATIONS

1. Wooden folding rule and steel tape - occasionally apply a few drops of light oil on the spring joints.
2. Steel tape - be careful not to kink or twist the steel tape as this could cause breakage.
3. Steel tape - wipe off excess moisture to keep from rusting.
4. Steel tape - do not use near exposed electrical parts.

SELF-CHECK REVIEW / PRACTICE QUESTIONS 9

1. The simplest and most common of all measuring tools in the construction industry is the steel rule.
    A.   True   B.   False

2. Why is a thin steel rule better than a thick one?
    A.   It's cheaper
    B.   Takes up less space in your tool box
    C.   More accurate measurement, because the division marks are closer to the work

PERFORMANCE / LABORATORY EXERCISE

Supplies needed:
    Steel rule
    Wooden folding rule
    Steel measuring tape
    Sharpened pencil

Try this:
    Measure various items around the room.

Check to be sure that:
    The hook is placed securely over the edge of the material being measured.

11.0.0 BENCH VISES

Vises are holding tools. They permit one worker to do work that would otherwise require two people.

11.1.0 DESCRIPTION AND SELECTION

The bench vise (*Figure 28*) has two sets of jaws: one to hold flat work and another to hold round work, such as piping. Some bench vises have swivel bases so that the vise can be turned in any horizontal position.

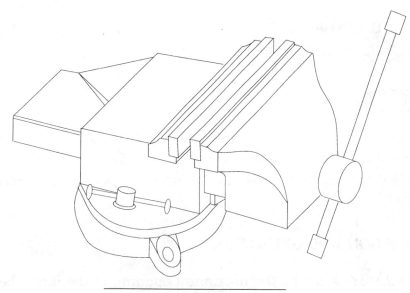

Figure 28. The Bench Vise

11.2.0 HOW TO USE A BENCH VISE

Step 1   Turn the sliding T-handle screw clockwise to clamp the object.

Note   Never hammer the handle tight or use a piece of pipe for leverage. This may damage the vise as well as the object being clamped.

11.3.0 SAFETY AND MAINTENANCE CONSIDERATIONS

1. Fasten vise securely to the bench.
2. Do not use the jaws of the bench as an **anvil**.
3. Saw as close to the jaws as possible.
4. Support the ends of long stock being held in vise.
5. Clamp work evenly in the vise.
6. Keep threaded parts clean.

1. The bench vise has_____set(s) of jaws.

    A. 1   B. 2

2. One set of vise jaws are for holding flat work; the other for holding_____work.

    A. Slippery   B. Excess   C. Hot   D. Round

3. It is OK to use the jaws of the vise as an anvil._____

    A. True   B. False

12.0.0   C-CLAMPS

Many types of clamps are manufactured, each designed to solve a different holding problem. C-clamps are probably the most widely used. Named for the shape of the clamp frame, C-clamps are made in various sizes and come in handy for construction workers.

12.1.0   DESCRIPTION AND SELECTION

C-clamps *(Figure 29)* are sized by the maximum opening of the jaw. They are made in sizes from 1" to 24". The depth, or throat, of the clamp is also important, as it determines how far from the edge of the work the clamp can be attached. Regular C-clamps have throat depths from 1" to 16 ".

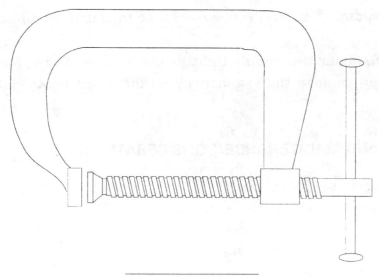

Figure 29.   C-Clamp

CORE CURRICULA TRAINEE TASK MODULE 00103

12.2.0 HOW TO USE A C-CLAMP

Step 1   For wood and other soft materials being clamped, place pads or thin blocks of wood between the work piece and the clamp anvils to protect the work.

Step 2   Tighten the clamp's T-bar handle only hand-tight.

Note   Do not use pliers or a section of a pipe on the handle.

If you are clamping work that has been glued, be certain not to tighten the clamps so much that all the glue is squeezed out of the joint.

12.3.0 SAFETY AND MAINTENANCE CONSIDERATIONS

1.   Store C-clamps by clamping them in a rack.
2.   Use pads or thin wood blocks to protect work.
3.   Get rid of clamps that have bent frames.
4.   Do not over-tighten.
5.   Clean and oil threads.
6.   Check the swivel at the end of the screw. Make sure it turns freely.
7.   Never use a C-clamp for hoisting (pulling up) work.

SELF-CHECK REVIEW / PRACTICE QUESTIONS 11

1.   The sizes of a C-clamp are determined by:_____

A.   The width of the metal
B.   The maximum opening of the jaws
C.   The length of the T-handle screw

2.   It is best to place_____between the work and the clamps to protect the work.

A.   A screwdriver   B.   A hammer   C .   Pads or thin wood blocks

3.   Tighten the T-bar hand tight and then force it tighter with a wrench._____

A.   True   B.   False

The right saw for the job makes the cutting easy. Here is a typical handsaw and its parts:

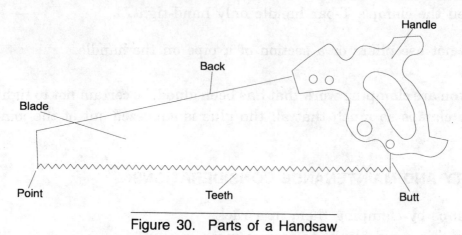

Figure 30. Parts of a Handsaw

13.1.0 DESCRIPTION AND SELECTION

The handsaw's blade *(Figure 30)* is made of tempered steel to hold sharpness and resist bending and buckling. A key feature of any saw blade is the number, shape, size, slant and direction of the teeth. Saw teeth are set or angled alternately in opposite directions to make a cut (or **kerf**) slightly wider than the thickness of the saw blade itself.

Generally, the fewer the teeth per inch (called **points**), the coarser and faster the cut. More teeth per inch means a slower but smoother cut surface. Ripping saws, which cut along the grain of the wood, generally have 5 to 6 points per inch. Backsaws have 12 to 14 points per inch, for a smooth, close-fitting cut surface.

Two often-used hand saws are the **crosscut saw** and the **rip saw.**

The crosscut saw is designed to cut across the grain of wood. The teeth are like tiny knives, with 12 to 14 points per inch. Blade lengths vary from 20" to 28". For most general uses, a 24" or 26" saw is a good length *(See Figure 31)*.

The rip saw cuts along the grain of the wood. It has 5 to 6 points per inch. This type of saw, which cuts along the grain, meets less resistance that a saw cutting across the grain.

SAW TYPE	POINTS	CUTTING STYLE
Crosscut Saw	12-14	Slower but smoother
Rip Saw	5-6	Faster but coarser

Figure 31. How the Saws Cut

13.2.0 HOW TO USE A CROSSCUT SAW

Step 1   Mark the cut to be made with a square or other measuring tool.

Step 2   Make sure the piece to be cut is well supported – on a sawhorse, jack, or other support.

Support the scrap end as well as the main part of the wood to prevent it from splitting as the saw kerf nears the edge. With smaller stock, you can support the scrap end of the piece with your free hand.

Step 3   Place the saw teeth on the edge of the board, just at the outside edge of the mark.

Step 4   Use the part of the blade nearest the handle end of the saw, as your first stroke will be pulled toward your body.

Step 5   Use the thumb of your left hand (hand not sawing) to support the saw vertical to the work.

Step 6   With the saw at about a 45-degree angle to the wood, pull the saw to make a small groove.

Step 7   Start sawing slowly, increasing the length of the stroke as the kerf deepens.

Step 8   Continue to saw with the blade at a 45-degree angle to the board.

Note   Don't push or "ride" the saw into the wood. Let the weight of the saw set the rate of cutting. It's easier to control the saw and is less tiring that way.

13.3.0   HOW TO USE A RIP SAW

Step 1   Mark and start a ripping cut the same way you'd make a start using a crosscut saw.

Step 2   Once you've started the kerf, saw with the blade at a steeper angle to the wood – about 60 degrees.

Note   If the saw starts to wander from the line, angle the blade toward the desired line cut.

If the saw blade binds in the kerf, wedge a thin piece of wood into the cut.

13.4.0   SAFETY AND MAINTENANCE CONSIDERATIONS

1. Rust is deadly for the saw blade. If it starts to rust, remove the rust with a fine emery cloth and apply a coat of light machine oil.
2. Always lay a saw down gently. Never let saw teeth come in contact with stone, concrete, or metal.
3. Have your saw sharpened by an experienced sharpener.
4. Keep a stable stance so you are not thrown off balance on the last stroke.

SELF-CHECK REVIEW / PRACTICE QUESTIONS 12

1. Fill in the parts of the handsaw *(Figure 32)* with the correct term from the following: Blade, Point, Back, Teeth, Handle, Butt

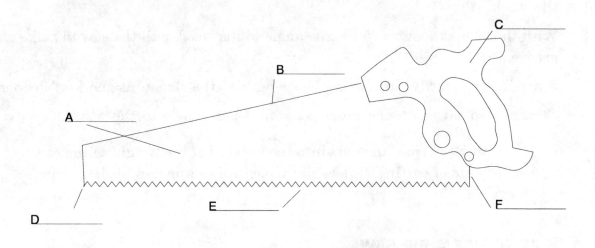

Figure 32.   Fill in the Parts of the Handsaw

2. Points refers to:_____
    A.   The teeth of a saw
    B.   The type of wood on a saw
    C.   Teeth per inch on a saw

3. The smaller the point, the coarser and faster the cut._____
    A.   True   B.   False

4. Ripping saws have_____points per inch; crosscuts have_____.
    A.   12 to 14   B.   5 to 6

5. Which saw makes a coarse cut?_____
    A.   Ripping   B.   Crosscut

PERFORMANCE / LABORATORY EXERCISE

Supplies needed:
  Bench vise
  Crosscut saw
  Rip saw
  Wood scrap at least 15" long
  Combination square.

Try this:
  1. Use the crosscut saw to cut across the grain to cut a piece of wood that is 12" long.
  2. Use the rip saw to rip wood with the grain to cut the piece of wood you just cut in half.

Check to be sure that:
  1. You first mark the place where the cut is to be made.
  2. You support the piece to be cut.
  3. You pull the first stroke towards your body.
  4. The saw should be vertical to the work.

14.0.0 FILES

Files are used for cutting, smoothing, or shaping metal parts. They can also be used for finishing and shaping all metals, except hardened steel, and for sharpening many other tools and cutting instruments.

14.1.0 DESCRIPTION AND SELECTION

Files are usually made from a hardened piece of high-grade steel with slanting rows of teeth. They are sized by the length of the body. For most practical sharpening jobs, files range from 4" to 14". A 10" or 12" inch mill bastard file (flat and fairly coarse teeth) is useful for sharpening shovels, hoes, and spades. It also removes metal quickly when you need to re-dress (sharpen) axes and hatchets.

The first step in filing is file selection. There's a specific type for each of the common soft metals, for the hard ones, for plastics, and for wood. In general, the teeth of files for soft materials are very sharp and widely spaced. Those for hard materials are closer and stockier. The shape of the teeth also differs according to the material to be worked. If you use a soft-material file on hard material, the teeth quickly chip and dull. If you use a hard-material file on soft material, the teeth clog.

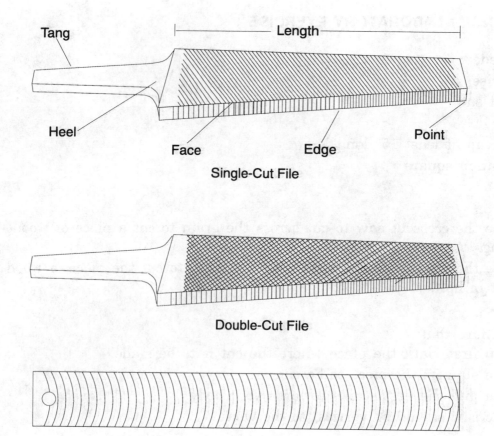

Single-Cut File

Double-Cut File

Curved-Tooth File

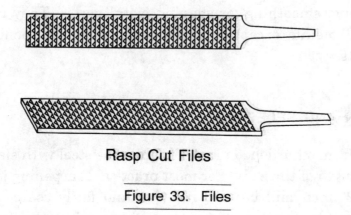

Rasp Cut Files

Figure 33.   Files

Files *(Figure 33)* are classified by the cut of the teeth.   Classifications include:

- single-cut and double-cut
- rasp-cut
- curved-tooth

Here is how you would use each type *(Figure 34)*:

Type:	**Description:**	**When Used:**
Rasp-cut file	The teeth are individually cut, not connected to each other.	This file produces an extremely rough surface; used mostly on aluminum, lead, and other soft metals for removing waste materials. Also used on wood and horses hooves.
Curved-tooth file	Single row of widely spaced teeth cut on the blade to aid in self-cleaning.	Used on aluminum and steel sheets to provide a flat, smooth finish.
Single-cut file	Single, straight-edged teeth running across it at an angle.	For a smooth, keen surface, e.g., sharpening rotary mower blade.
Double-cut file	Second set of teeth criss-crossing the first ones. Types are: • bastard • second cut • smooth	Fast cutting action. Removes material quickly. Gives rough finish.

Figure 34. Type and Uses of Files

14.2.0 HOW TO USE A FILE

Step 1   Mount the work to be filed in a vise at about elbow height.

Step 2   Do not hunch directly over your work. Stand back from the vise a little with your feet about 24" apart, the right foot ahead of the left*.

Step 3   Most files have handle attachments. Always put on a handle before using the tool to avoid injuring your hand.

Step 4   Hold the file with the handle in your right hand, the tip of the blade in your left*.

Step 5   For average work, the tip is best held with the thumb on top of the blade, the first two fingers under it. For heavy work, use a full hand grip on the tip.

Step 6   Apply pressure only on the forward stroke.

Step 7   Raise the file from the work on the return stroke to prevent file damage.

Step 8   Keep the file flat on the work. Clean it by tapping lightly at the end of each stroke.

* Switch if left-handed.

14.3.0 SAFETY AND MAINTENANCE CONSIDERATIONS

1. Most files have handle attachments. Always put on a handle before using the tool to avoid injuring your hand.
2. Filings should be brushed from between the teeth with a wire brush, pushed in the same direction as the line of the teeth.
3. Always use the correct file for the material being worked.
4. Do not let the material vibrate in the vise. It dulls the file teeth.
5. Store files in a dry place and keep them separated so that one will not chip or damage another.

SELF-CHECK REVIEW / PRACTICE QUESTIONS 13

1. Files are used in the construction trade primarily for:

    A. Removing paint
    B. Shining metal
    C. Cutting, smoothing, or shaping metal parts

2. Files are classified by:

    A. The cut of the teeth
    B. The type of handle

PERFORMANCE / LABORATORY EXERCISE

Supplies needed:

    One file of each type:
        Rasp-cut
        Curved-tooth
        Single-cut
    Scrap metal pieces (see instructor).

Try this:
1. Practice using the various files on the various metals.
2. Determine which type file is best for each type metal.
3. Verify that you are using the best file for the metal selected (*see Figure 34*).

Check to be sure that:
1. You are using the correct file for the type of metal being filed (*See Figure 34*).
2. The work is mounted at elbow height (approximately).
3. A handle has been attached to the file.
4. Pressure is applied only on the forward stroke.
5. The file is kept flat on the work.

15.0.0 CHISELS AND PUNCHES

A chisel is a metal tool with a sharpened, beveled edge that is used to cut and shape wood, stone, or metal. A chisel can cut any metal that is softer than the steel of the chisel. Two kinds of chisels will be addressed here: the **wood chisel** and the **cold chisel**. Both chisels are made from heat-treated steel to increase the hardness of their cutting blades.

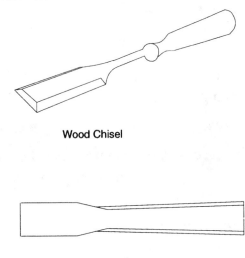

Wood Chisel

Cold Chisel

Figure 35. Wood and Cold Chisels

Punches are used to indent metal before drilling a hole, to drive pins, and to align holes in two mating parts. They are made of hardened and tempered tool steel and generally come in various sizes.

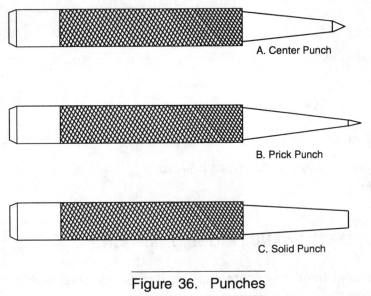

A. Center Punch

B. Prick Punch

C. Solid Punch

Figure 36. Punches

15.1.0 DESCRIPTION AND SELECTION

The **wood chisel** is used to make openings or notches in wooden structural material. A common use of a wood chisel is to make a recess for butt-type hinges.

Cold chisels are used to cut cold metal, hence, the name (*See Figure 35*).

Three types of **punches** are the **prick punch**, **center punch**, and **solid punch** (*See Figure 36*). The prick and center punches are used for making small locating points to be used for drilling holes. The solid punch is made for punching holes in light-gauge metals.

15.2.0 HOW TO USE A WOOD CHISEL

Step 1   Outline the recess to be chiseled.

Step 2   Set the chisel at one end of the outline, with its edge on the cross grain line and the bevel facing the recess to be made (*See Figure 37*).

Step 3   Strike the chisel head lightly with a **mallet**.

Step 4   Repeat this process at the other end of the outline, again with the bevel of the chisel blade toward the recess. Then make a series of cuts about 1/4 inch apart from one end of the recess to the other.

Step 5   To pare (trim) away the notched wood, hold the chisel bevel up to slice from the edge of the wood inward.

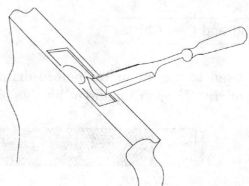

Figure 37.   Using a Wood Chisel

15.3.0 HOW TO USE A COLD CHISEL

Step 1   Hold the object to be cut in a vise if possible.

Step 2   Hit the opposite end or handle with a hammer. The chisel is forced into and through the material.

15.4.0 SAFETY AND MAINTENANCE CONSIDERATIONS

1.   Do not allow the chisel head to become mushroomed, or flattened. Always remove this excess metal by grinding or filing (*Figure 38*).

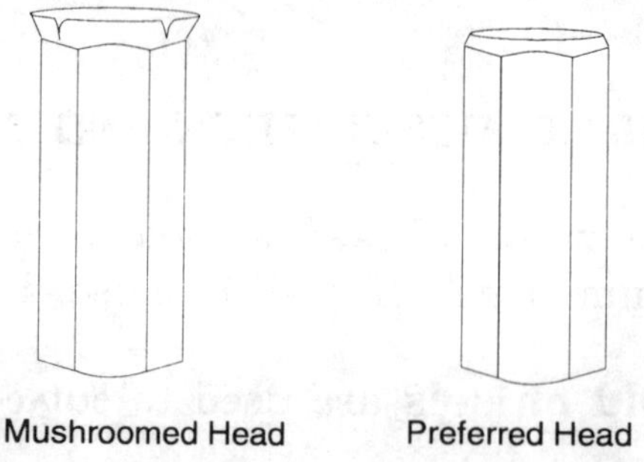

Figure 38.   Mushroomed and Preferred Chisel Heads

Mushroomed Head   Preferred Head

CAUTION   Serious injuries could result from metal fragments breaking off of mushroomed chisel heads.

2. Always wear safety goggles.
3. Never use cold chisels for cutting or splitting stone or concrete.
4. In order for a wood chisel to cut well, the blade needs to be beveled (sloped) at a precise 25-degree angle. The cutting edge must be sharpened on an oil stone to produce a keen edge. For a cold chisel, a 60-degree angle is required.

SELF-CHECK REVIEW / PRACTICE QUESTIONS 14

1. A chisel is used to cut and shape:_____

    A. Diamonds, emeralds, rubies
    B. Wood, teeth, fingernails
    C. Wood, stone, or metal

2. An example of the use for a wood chisel is to:_____

    A. Make a recess for butt type hinges, as in a door.
    B. Spread paint.
    C. Cut down a tree.

3. An example of the use for a solid punch is:_____

    A. Leveling a uneven piece of work.
    B. Punching holes in light gauge metals.
    C. Wedging a large piece of equipment.

4. A mushroomed-headed chisel is dangerous because the loose metal could end up in your or your partner's eye._____

    A. True   B. False

16.0.0   PLUMB BOB

The **plumb bob** *(Figure 39)* uses the force of gravity to make a vertical (plumb) line. Plumb bobs come in different weights, 12 oz., 8 oz., and 6 oz. being the most common.

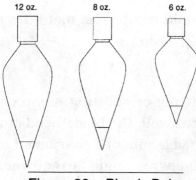

Figure 39. Plumb Bobs

16.1.0 DESCRIPTION AND SELECTION

A plumb bob consists of a pointed weight attached to a string. It uses the force of gravity to make a vertical (plumb) line. When the weight is suspended by the string and allowed to hang freely, the string is plumb (*See Figure 40*). Suppose you want to install a post under a beam. A plumb bob can indicate the point on the floor that is directly under the section of the beam you need to support.

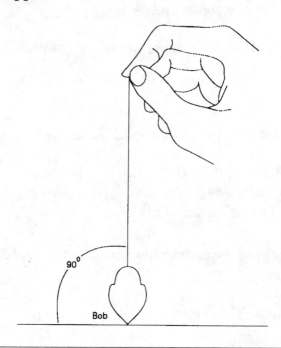

Figure 40. Finding True Vertical (Plumb) With a Plumb Line

16.2.0 HOW TO USE A PLUMB BOB

Step 1   The line attached to the plumb bob must be attached at the exact top center of the plumb bob.

Step 2   Hang the bob from a vertical member.

Step 3   When the weight is suspended by the string and allowed to hang freely, the string is plumb (vertical).

Note   When using the plumb bob outside, be aware that the wind may affect the true reading of the plumb bob.

16.3.0 SAFETY AND MAINTENANCE CONSIDERATIONS

1. Do not drop the plumb bob on its point. A bent or rounded point causes inaccurate readings.

SELF-CHECK REVIEW / PRACTICE QUESTIONS 15

1. Plumb refers to_____and level refers to_____. (Fill in the blanks with either horizontal or vertical.)

2. The line attached to the plumb bob must be attached at the exact top center of the plumb bob._____

   A. True   B. False

3. When using a plumb bob outside, you should remember that the_____may effect the plumb bob.

   A. Earth rotation
   B. Wind
   C. Gravity
   D. Point of suspension

PERFORMANCE / PRACTICE EXERCISE

Supplies needed:

   Plumb bob.

Try this:
   Use the plumb bob to check the plumb of various vertical objects in the room.

Check to be sure that:
   1. The line attached to the plumb bob is attached at the exact top center of the plumb bob.
   2. The bob is suspended freely.

Socket wrench sets include different combinations of **sockets** and **ratchet** (handles). The socket is the part that grips the nut or bolt.

17.1.0 DESCRIPTION AND SELECTION

Most common sockets have 6 or 12 gripping points. Sockets *(Figure 41)* also come in different lengths. The long socket is called a deep socket and is used when normal sockets will not reach down over the end of the bolt to grip the nut.

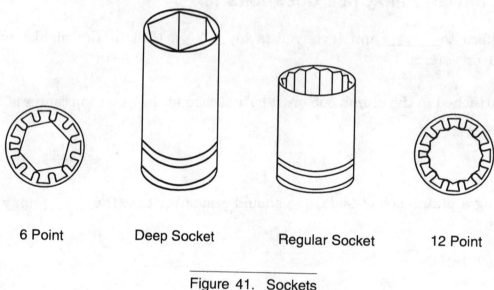

| 6 Point | Deep Socket | Regular Socket | 12 Point |

Figure 41. Sockets

Socket sets contain different types of handles for different uses. The ratchet ha*ndle, (Figure 42)* one type of socket handle, has a small lever on the handle used to change the turning direction.

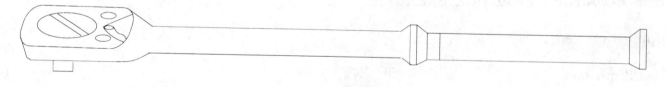

Figure 42. Ratchet Handle

17.2.0 HOW TO USE SOCKETS AND RACHETS

Step 1   To use a socket with a ratchet, select a socket that fits the fastener.

Step 2   Place the square end of the socket over the spring-loaded button on the ratchet shaft.

Step 3   Place the tool over the fastener.

Step 4   Pulling on the handle in one direction will turn the nut. Moving the handle in the other direction has no effect. To reverse the direction of the socket, use the adjustable lock mechanism.

17.3.0 SAFETY AND MAINTENANCE CONSIDERATIONS

1. Never force the ratchet handle beyond hand-tight. This could break the head of the fastener.
2. Never use a cheater pipe.

SELF-CHECK REVIEW / PRACTICE QUESTIONS 16

1. The socket is the part that grips the nut or bolt._____

    A. True   B. False

2. The ratchet handle is one type of socket handle._____

    A. True   B. False

3. The small lever on the ratchet handle is used to:_____

    A. Change the socket   B.   Change the turning direction

PERFORMANCE / LABORATORY EXERCISE

Instructor will demonstrate.

18.0.0   TORQUE WRENCHES

Torque wrenches measure resistance to turning. They are invaluable when installing fasteners that must be properly tightened in sequence without distorting the workpiece. Torque wrenches are only used when tightening procedures specify a torque setting for a bolt.

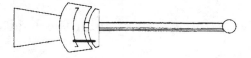

Figure 43. Torque Wrench

18.1.0 DESCRIPTION AND SELECTION

Torque specifications are usually stated in inch-pounds for smaller fasteners or foot-pounds for larger fasteners.

18.2.0 HOW TO USE A TORQUE WRENCH

Step 1 Find out how many inch-pounds or foot-pounds to torque to.

Step 2 Find out the sequence to torque (which fastener comes first, second, etc.).

Step 3 Place the wrench on the fastener. Place one hand on the head of the wrench at the location of the bolt to support the bolt and ensure proper alignment.

Step 4 Watch the torque indicator as you tighten the bolt.

18.3.0 SAFETY AND MAINTENANCE CONSIDERATIONS

In order to obtain accurate settings, all threaded fasteners must be clean and undamaged.

SELF-CHECK REVIEW / PRACTICE QUESTIONS 17

1. Torque wrenches measure:_____

   A. Plumb   B. Level   C. Resistance to turning

2. Torque specifications are usually stated in_____for smaller fasteners or_____for larger fasteners. (Fill in the blank with either **foot-pounds** or **inch-pounds**).

19.0.0   WEDGES

A wedge is a piece of hard rubber, plastic, wood, or steel tapered to a thin edge. It is used to adjust elevation, tighten formwork, etc.

19.1.0 DESCRIPTION AND SELECTION

Proper wedging entails a wedging force greater than the load ultimately to be lifted.

19.2.0 HOW TO USE A WEDGE

Step 1   Position wedge at edge of load to be lifted/separated.

Step 2   Drive wedge under load with hammer.

19.3.0 SAFETY AND MAINTENANCE CONSIDERATIONS

1. Wear safety glasses and face shield when using a wedge, as a piece of the wedge could fly off.
2. Keep hands away from hammer side of wedge.

SELF-CHECK REVIEW / PRACTICE QUESTIONS 18

1. A wedge can only be made from wood.

    A. True   B. False

2. A wedge is used for:

    A. Opening a paint can
    B. Checking for straightness
    C. Adjusting elevation

20.0.0   CHALK LINES

A chalk line *(Figure 44)* is a piece of string or cord that has been coated with chalk. The line is stretched tightly between two points that are to be joined by a straight line and then "snapped" to transfer a chalky line to the surface. You can use a piece of string rubbed with chalk, if you only need to snap a couple of lines. But for steady use, a mechanical self-chalking line is much handier.

20.1.0 DESCRIPTION AND SELECTION

A mechanical self-chalking is made of a metal box containing a line on a reel. The box is filled with colored chalk powder. The line is automatically chalked each time it is withdrawn from the box. Some models have a point on the end of the box that can double as a plumb bob.

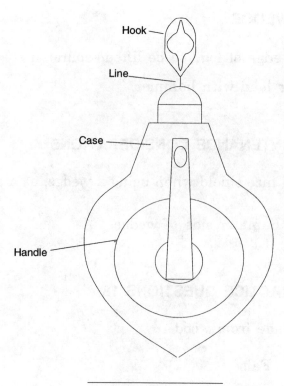

Hook —

Line —

Case —

Handle —

Figure 44. Chalk line

20.2.0 HOW TO USE A CHALK LINE

Step 1   Pull the line from the case. Have a partner hold one end. Stretch it between two points to be connected.

Step 2   After the line has been pulled tight, pull it up and then release it (snap it). This marks the surface underneath with a straight line of chalk.

20.3.0 SAFETY AND MAINTENANCE CONSIDERATIONS

The chalk line should always be kept in a dry place.

21.0.0 UTILITY KNIVES

A utility knife is especially useful for cutting roofing felt, asbestos shingles, vinyl and linoneum floor tiles, fiberboards, and sheetrock wall tiles. You can also use a utility knife to trim insulation.

21.1.0 DESCRIPTION AND SELECTION

This knife has replaceable razor-like blades and a cast aluminium handle about six inches long to hold the changeable blade. The handle is made in two halves, held together with a screw *(See Figure 45)*.

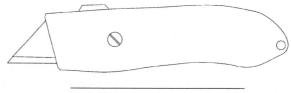

Figure 45. The Utility knife

The blade can be locked in the handle in one, two, or three positions, depending on the type of utility knife used. Some models feature a retractable blade; this kind is safer. Some have substitute blades for sawing or scraping.

21.2.0 HOW TO USE A UTILITY KNIFE

Step 1   Unlock the knife blade. Extend the knife out.

Step 2   Lock the blade in the out position.

Step 3   Place a scrap object, such as a piece of wood, under the object to be cut. This will protect the surface beneath the object.

Step 4   Use the razor-sharp side of the blade (the longer side) to cut straight lines.

Step 5   Unlock the blade; retract to the closed position. Lock the blade.

21.3.0 SAFETY AND MAINTENANCE CONSIDERATIONS

1. You can sharpen a utility knife blade, but the replaceable blades are so inexpensive that it makes more sense to replace it with a new one.
2. Always keep the blade closed and locked when not in use.
3. NEVER use a utility knife on live electrical wires. Death could result.

1. A utility knife can be used for cutting everything EXCEPT:_____

    A. Roofing felt
    B. Asbestos shingles
    C. Vinyl and linoleum floor tiles
    D. Fiberboards
    E. Sheetrock wall tiles
    F. Insulation
    G. Live electrical wires

2. How should the blade be kept when not in use?

    A. Open   B. Closed

22.0.0   CHAIN FALLS AND COME ALONGS

Chain falls and come alongs are useful lifting devices.

22.1.0 DESCRIPTION AND SELECTION

The chain fall *(Figure 46)* has an automatic brake that holds the load after it is lifted. As the load is lifted, a screw forces fiber discs together to keep the load from slipping. If the load increases, the brake pressure also increases. The brake holds the load until the lowering chain is pulled. Manual chain falls are operated by hand. There is also an electrical chain fall that is operated from an electrical control box. We will only be discussing the manual chain fall here.   Each identified part of the chain fall will be addressed following the illustration.

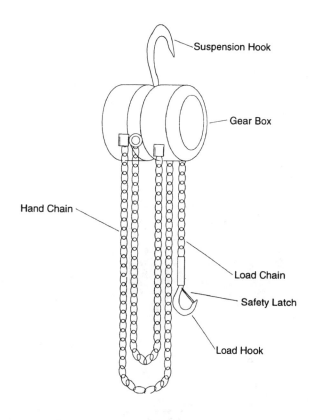

Figure 46. The Manual Chain Fall

SUSPENSION HOOK	A steel hook used to hang the chain fall. The suspension hook is one size larger than the load hook.
GEAR BOX	Contains the gears used for lifting power.
HAND CHAIN	A continuous chain used for operating the gear box.
LOAD CHAIN	The chain attached to the load hook and used for lifting loads.
LOAD HOOK	The hook attached to the load to be lifted.
SAFETY LATCH	Prevents load from slipping off hook.

A come along is a lifting device with a ratchet handle *(See Figure 47)*. Come alongs are used for short pulls on heavy loads. Come alongs are available in 1 to 6 ton capacities. Some come alongs have a chain for pulling; others use wire ropes. The parts of the come along are addressed following the illustration.

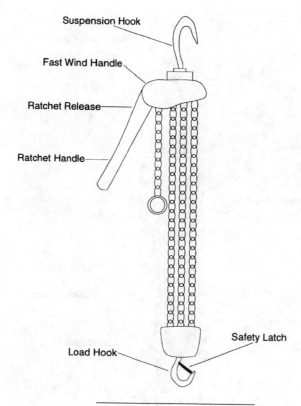

Suspension Hook

Fast Wind Handle

Ratchet Release

Ratchet Handle

Safety Latch

Load Hook

Figure 47. Come Along

FAST WIND HANDLE	Takes up or lets out the chain without using the ratchet handle.
RATCHET HANDLE	Operates the ratchet used to take up the chain.
RATCHET RELEASE	Releases the ratchet so the chain can be pulled out.

22.2.0 HOW TO USE A CHAIN FALL

Step 1   Use the load hook to hook to the load.

Step 2   Pull on the hand chain. One side lifts the load; the other will lower the load.

22.3.0 HOW TO USE A COME ALONG

Step 1   Use the load hook to hook to the load.

Step 2   Flip the ratchet release up to lift the load; down to lower the load.

Step 3   Use the ratchet handle like a car jack to either lift or lower the load.

Step 4   Use the fast wind ratchet release to take up slack (do not use the fast wind for lifting or lowering).

22.4.0 SAFETY AND MAINTENANCE CONSIDERATIONS

1. Lubricate per manufacturer recommendations. Do not get lubricant on clutches.
2. Store in proper place.
3. Inspect for wear.
4. Check proper operability on small load.
5. Make sure the support rigging is of adequate capacity to handle load.
6. Never stand under a load.
7. Keep hands away from pinch-points of chain.

SELF-CHECK REVIEW / PRACTICE QUESTIONS 20

1. Chain falls and come alongs are tools used for:_____

   A. Welding   B. Sawing   C. Lifting

2. It is important to be sure the support rigging is of adequate capacity to handle load. _____

   A. True   B. False

23.0.0   WIRE BRUSHES

A wire brush can be used for any application that requires cleaning of metal parts.

23.1.0 DESCRIPTION AND SELECTION

This wood-handled brush with steel bristles comes in handy for cleaning rusty tools, removing paint, cleaning **welds,** and other heavy duty uses.

23.2.0 HOW TO USE A WIRE BRUSH

Step 1   Use a wire brush for any of the purposes listed above.

CAUTION   Do NOT use a wire brush for finishing work. It will scratch the surface.

23.3.0 SAFETY AND MAINTENANCE CONSIDERATIONS

1. The metal bristles of wire brushes are subject to rust unless the tool is thoroughly dried before storage.
2. Do not use the wire brush for finishing work. It will scratch the surface.

SELF-CHECK REVIEW / PRACTICE QUESTIONS 21

1. A wire brush can be used for applications that require cleaning of metal parts._____

    A. True   B. False

2. It is OK to use a wire brush for finishing work._____

    A. True   B. False

PERFORMANCE / LABORATORY EXERCISES

A.   HAMMER AND NAILS

Supplies:

For this exercise, you will need the following: Claw hammer, 2 nails, scrap piece of wood.

Steps:

1. Hammer 2 nails into the board.

2. Remove 1 of the nails.

B.   SCREWDRIVERS AND SCREWS

Supplies:

For this exercise, you will need the following: Straight blade screwdriver, 1 straight blade screw, Phillips head screwdriver, 1 Phillips head screw, 1 piece of "soft" wood.

Steps:

1. Use the straight blade screwdriver to drive in a straight blade screw.

2. Use the Phillips head screwdriver to drive in a Phillips head screw.

3. Remove (unscrew) 1 of each type.

C.   WRENCH AND BOLTS

Supplies:

For this exercise, you will need the following: Combination wrenches (varying sizes), adjustable wrenches (varying sizes), instructor-prepared board with varying size bolts.

Steps:

1. Remove 3 varying sized nuts or bolts.

2. Practice reinstalling them.

D. WIRE CUTTING

Supplies:

For this exercise, you will need the following: Lineman pliers, 1 piece of wire,

Steps:

1. Cut the wire into two pieces.

E. CHECKING PLUMB AND LEVEL

Supplies:

For this exercise, you will need the following: Spirit level, 2 objects (provided by instructor)

Steps:

1. Use the spirit level to check the plumb of the objects provided by the instructor.

F. WOOD CUT

Supplies:

For this exercise, you will need the following: Bench vise, crosscut saw, rip saw, wood scrap (at least 15" long), combination square.

Steps:

1. Use the crosscut saw to cut a 8" long piece from a wood panel.

APPENDIX A: Answers to SELF-CHECK REVIEW / PRACTICE QUESTIONS

SELF-CHECK REVIEW 1

1. C
2. A
3. A
4. A
5. A

SELF-CHECK REVIEW 2

1. D
2. D
3. B
4. B

SELF-CHECK REVIEW 3

1. B
2. C
3. B

SELF-CHECK REVIEW 4

1. B
2. B
3. C

SELF-CHECK REVIEW 5

1A. 3
1B. 1
1C. 2
2. B
3. A
4. B

SELF-CHECK REVIEW 6

1. B
2. A
3. B
4. A
5A. 2
5B. 3
5C. 1

SELF-CHECK REVIEW 7

1. Vertical; Horizontal
2. B
3. A
4. C

SELF-CHECK REVIEW 8

1. C
2. C
3. B

SELF-CHECK REVIEW 9

1. A
2. C

SELF-CHECK REVIEW 10

1. B
2. D
3. B

SELF-CHECK REVIEW 11

1. B
2. C
3. B

SELF-CHECK REVIEW 12

1A. Blade
1B. Back
1C. Handle
1D. Point
1E. Teeth
1F. Butt
2. C
3. B
4. B; A
5. A

SELF-CHECK REVIEW 13

1. C
2. A

SELF-CHECK REVIEW 14

1. C
2. A
3. B
4. A

SELF-CHECK REVIEW 15

1. vertical; horizontal
2. A
3. B

SELF-CHECK REVIEW 16

1. A
2. A
3. B

SELF-CHECK REVIEW 17

1. C
2. inch-pounds; foot-pounds

SELF-CHECK REVIEW 18

1. B
2. C

SELF-CHECK REVIEW 19

1. G
2. B

SELF-CHECK REVIEW 20

1. C
2. A

SELF-CHECK REVIEW 21

1. A
2. B

The NCCER makes every effort to keep these manuals up-to-date and free of technical errors. We appreciate your help in this process. If you have an idea for improving this manual, or if you find an error, a typographical mistake, or an inaccuracy in the *Wheels of Learning*, please write us, using this form or a photocopy. Be sure to include the exact module number, page number, a description of the problem, and the correction, if possible. We'll do our best to correct it in later editions. Thank you for your assistance.

Write: *Wheels of Learning*
National Center for Construction Education and Research
P.O. Box 141104
Gainesville, FL 32614-1104

Fax: 352-334-0932

WHEELS OF LEARNING USER UPDATE

Please let us know if you have found an inaccuracy, error, or other problem in a *Wheels of Learning* manual. Use this form or write us a letter. Please be sure to tell us the exact module name and module number, the page number, and the problem. Thanks for your help.

Craft _____ Module Name _____

Module Number _____ Page Number(s) _____

Description of Problem _____

_____

_____

(Optional) Correction of Problem _____

_____

_____

(Optional) Your Name and Address _____

_____

Introduction to Power Tools

Module 00104

Core Curricula Trainee Task Module 00104

NATIONAL
CENTER FOR
CONSTRUCTION
EDUCATION AND
RESEARCH

INTRODUCTION TO POWER TOOLS

Objectives

Upon completion of this module, the trainee will be able to:

1. Identify commonly used power tools of the construction trade.
2. Recognize safe use of power tools.
3. Explain the procedures to properly maintain these power tools.

Prerequisites

Successful completion of the following Task Modules is required before beginning study of this Task Module: NCCER Task Module 00101, *Basic Safety;* NCCER Task Module 00102, *Basic Math;* NCCER Task Module 00103, *Introduction to Hand Tools.*

WARNING!   Because trainees will be operating electrical power tools during this module, it is *required* that the *Basic Safety* Module be successfully completed before going on. **Additionally, wear appropriate personal protective equipment while operating or near the operation of any power tool — if in doubt, check with your instructor or supervisor.**

How To Use This Manual

During the course of completing this module, you will be taught and will practice the safe use of the most commonly used basic power tools. *Self-Check Review / Practice Questions and Performance / Laboratory Exercises* will follow the introduction of most tools. The answers to these written exercises are found in Appendix A of this manual, titled *Self-Check Review / Practice Questions and Performance / Laboratory Exercises.* Your instructor will observe the performance of the *Performance / Laboratory Exercises* to ensure proper tool use.

New terms will be introduced in **bold** print. The definition of these terms can be found in the front of this manual, under *Trade Terms Introduced in This Module.*

Required Student Materials

Equipment

1. Drills
   A. Electric drill
   B. Cordless drill
   C. Pneumatic drill
   D. Electro-magnetic drill

2. Saws
   A. Circular saw
   B. Saber saw
   C. Reciprocating saw
   D. Portable, hand-held band saw
   E. Portable jig saw

3. Sanders and Grinders
   A. Angle grinder
   B. End grinder
   C. Bench grinder

4. Miscellaneous Tools
   A. Jack hammer (paving hammer)
   B. Porta-Power (hydraulic jack)
   C. Powder actuated (explosive) tools

Materials

1. Student Manual
2. Safety gloves
3. Sharpened pencil
4. Scrap linoleum tiles
5. Scrap metal of various thicknesses
6. Protective eye gear
7. Ear plugs
8. Scraps of wood (some hard, some soft)
9. Concrete block
10. Spark deflector

Course Map Information

This course map shows all of the *Wheels of Learning* task modules in the Core Curricula. The suggested training order begins at the bottom and proceeds up. Skill levels increase as a trainee advances on the course map. The training order may be adjusted by the local Training Program Sponsor.

Course Map: Core Curricula, Introduction To Power Tools

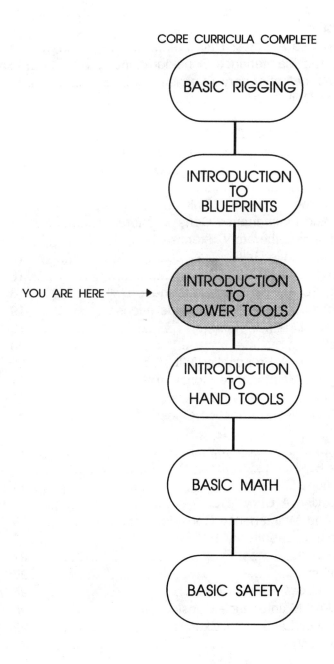

CORE CURRICULA COMPLETE

BASIC RIGGING

INTRODUCTION TO BLUEPRINTS

YOU ARE HERE ⟶ INTRODUCTION TO POWER TOOLS

INTRODUCTION TO HAND TOOLS

BASIC MATH

BASIC SAFETY

TABLE OF CONTENTS

TABLE OF CONTENTS (continued)

Trade Terms Introduced In This Module

Abrasive: Substance used to wear away material (e.g., sandpaper).

AC: Alternating current (wall plug).

Auger: A tool with a spiral cutting edge for boring holes in wood, etc.

Bevel cutting: Cutting at an angle other than a right angle; sloped.

Chuck: A clamping device which holds the attachment to be turned; for example, the chuck of the drill holds the drill bit.

Chuck key: Small T-shaped steel pieces used to open and close the chuck on power drills.

Combustible: Capable of catching fire and burning.

DC: Direct current (battery).

Diameter: A straight line passing through the center of a circle, from one side to another. A drilling or boring tool used in a drill.

Electro-magnetic: A soft iron core surrounded by a coil of wire that temporarily becomes a magnet when an electric current flows through the wire. An electro-magnetic drill is referred to in this course.

Fastener: A device used to attach or secure materials of various types to each other, such as clasp, hook, lock, etc.

Ferromagnetic: Of or characteristic of substances, as iron, nickle, cobalt, and various alloys, that have magnetic properties.

Grit: The texture produced by any of several sandstones with large, sharp grains, often used for grindstones.

Ground Fault Protection: A safety device that cuts power as soon as it senses any imbalance between incoming and outgoing current.

Hazardous Materials: Materials (e.g., chemicals) which must be transported, stored, applied, handled, and identified in accordance with Federal, State, or Local regulations. Must be accompanied by Material Safety Data Sheets (MSDS).

Hydraulic tools:	Tools powered by fluid pressure. Hand pumps or electric pumps are used to produce hydraulic pressure.
Kerf:	Cut or channel made by a saw.
Masonry:	Building material made of stone, brick, concrete block, etc.
Perpendicular:	At right angles.
Pneumatic tools:	Air-powered tools. Electric-or gasoline-powered compressors produce the pneumatic pressure.
Reciprocate:	To move alternately back and forth.
R.P.M.:	Revolutions per minute.
Shank:	The smooth portion of a drill bit that fits into the chuck.
Studs:	Upright pieces in the outer or inner walls of a building to which panels, siding, etc. are nailed.
Tempered:	A heat treatment used to create or restore hardness to steel.
Trigger lock:	A small lever or part which when pulled or pressed release or activates a locking catch or spring.

1.0.0 INTRODUCTION

Welcome to the next phase in your journey towards becoming a craftworker. Using power tools will open up a new world of possibilities in working in your craft area.

This introductory section presents a short overview of how the various types of power tools are powered and presents a word of caution concerning safety.

1.1.0   ELECTRIC, PNEUMATIC, HYDRAULIC, AND POWDER (EXPLOSIVE) TOOLS

Tools can be powered in one of four ways: electric, pneumatic, hydraulic, or powder (explosive).

Electric-powered tools are tools that get their power through electricity. They must be operated from either an **AC** (wall plug) or **DC** (battery) source.

Pneumatic tools are air-powered tools. Electric or fuel-powered compressors often produce the pneumatic pressure.

Hydraulic tools are powered by fluid pressure. Hand pumps or electric pumps are used to produce the hydraulic pressure.

Powder-actuated tools use energy from a gunpowder charge to drive a **fastener** into steel or concrete.

1.2.0   SAFETY

A prerequisite of this course is the completion of the Safety Module.

WARNING!	IF YOU HAVE NOT TAKEN THE SAFETY MOD-ULE, STOP HERE! YOU MUST TAKE THE SAFETY MODULE TO CONTINUE. Additionally, appropriate personal protective equipment must be worn while operating or near the operation of any power tool. If in doubt ask your instructor or supervisor.

Take the Safety Module before continuing with this module. There is a great potential for accidents in using power tools incorrectly or without knowing the safety requirements that apply to each. Though safety issues relating to each tool will be covered in this module, the more general safety issues relating to electricity, work area, safety equipment, etc., which were addressed in the Safety Module, are not addressed here. This information is vital for working with power tools.

WARNING!	*Always* disconnect the power source for any tool *before* replacing bits, blades, discs, etc. *Always* disconnect the power source *before* performing maintenance on any power tool. Never use trigger locks on any power tool.

1.3.0   SELF-CHECK REVIEW / PRACTICE QUESTIONS 1

1. The four ways that power tools can be powered are:
   A.   Electric, gas, pneumatic, powder
   B.   Electric, pneumatic, muscle, powder
   C.   Electric, pneumatic, hydraulic, powder
   D.   Electric, systematic, hydraulic, powder

2. How do electric tools get their power?
   - A.   Carbon dioxide
   - B.   AC (wall plug) or DC (battery) source
   - C.   Fluid pressure
   - D.   Radiowave
3. How do pneumatic tools get their power?
   - A. Air pressure
   - B. Salt water
   - C. Gunpowder
   - D. Gasoline

4. How do hydraulic tools get their power?
   - A. Gunpowder
   - B. Rotary engine
   - C. Air pressure
   - D. Fluid pressure

5. How do powder-actuated tools get their power?
   - A. Talcum powder
   - B. Baking powder
   - C. Baby powder
   - D. Gunpowder

6. It is okay to continue this course if you have not taken the Safety Module.
   - A. True
   - B. False

7. It is important that you wear appropriate personal protective equipment when operating or near the operation of any power tool?
   - A. True
   - B. False

2.0.0   DRILLS (POWER)

The power drill is an important and frequently used tool. The most common use of this tool is to make holes by driving **drill bits** into wood, metal, plastic and other materials. However, with a variety of attachments and accessories, the power drill can also serve as a sander, polisher, grinder - even as a saw.

2.1.0 POWER DRILLS (IN GENERAL)

In this section, we will address various types of power drills including:

- Electric Drill
- Cordless Drill
- Hammer Drill
- **Electro-magnetic** Drill
- Pneumatic Drill (air hammer)

Since most of these drills are similar in their general function and description, common traits will be discussed first.

2.1.1 Description

Most power drills have a pistol grip with a trigger switch for controlling power. The harder you pull on the trigger, the faster the speed.

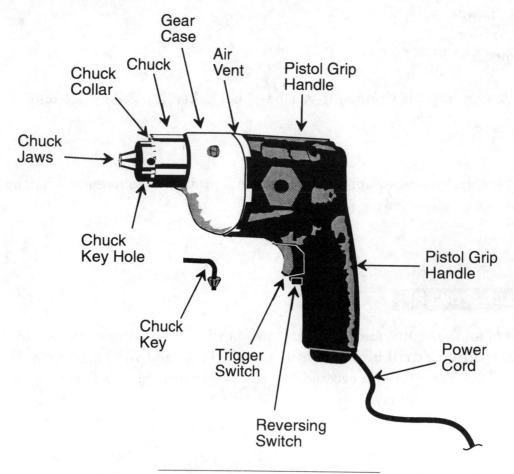

Figure 1. Parts of the Power Drill

What would you do if you were drilling and the drill bit got stuck in the material? Switch the direction of the reversing switch (*see Figure 1*). This allows you to back the drill out of the material.

Most drills have replaceable bits for use on various types of jobs. *Figure 2* shows different types of bits.

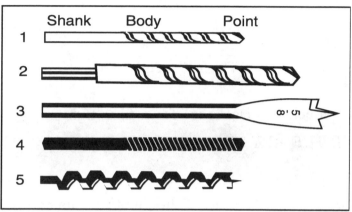

Figure 2. Drill Bits

Bits 1 and 2 are used to drill wood and plastics at high speeds or metal at a lower speed. Bit 3 is a flat-bladed spade bit, ranging in size from 1/2" to 1-1/2". They are used in wood only.

Bit 4 bit is a **masonry** bit. These are used in concrete, stone, slate, and ceramic. Bit 5 is the **auger** drill bit. This bit is used for drilling wood and other soft materials, but *not* for drilling metal.

As a rule, the point of a bit should be sharper for soft materials than for harder ones.

2.1.2 How To Use A Power Drill

Step 1   Wear appropriate personal protective equipment.

Step 2   To load the bit in the drill:

A. Disconnect the power source. Open the **chuck** (turn its outer barrel clockwise) until the bit **shank** can be inserted.

B. Tighten the chuck by hand until the jaws grip the bit. Be sure to keep the bit centered as you tighten it.

C. Insert the **chuck key** (*see Figure 3*) in one of the holes in the chuck so that the key's gear meshes with the geared end of the chuck. In larger drills, tighten *all* the holes in the 3-jawed chuck.

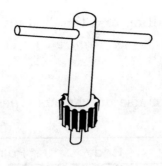

Figure 3. Chuck Key

D. Turn the chuck key clockwise for a firm tightening.

E. REMOVE THE KEY FROM THE CHUCK.

CAUTION   *Do not forget* to remove the key from the chuck, as the key could fly out and cause injury when you start the drill.

Step 3   Make an indent precisely where you want the hole drilled.

A. In wood, use a small punch to make an indent.

B. In metal, use a center punch.

Step 4   Firmly clamp or support the work being drilled.

Step 5   With the drill **perpendicular** to the material surface, start the drill motor. Be sure the drill is rotating in the proper direction (with bit facing *away* from you, it should be turning clockwise). Apply only moderate pressure when drilling. The drill motor should operate close to full revolutions per minute (**r.p.m.**).

Note   If the drill bit gets stuck in the material while drilling, release the trigger switch, switch the direction of the reversing switch and turn back on. This allows you to back out the drill. Switch it back to the correct turning position when finished backing out.

Step 6   Relax the pressure when the bit is about to emerge from the opposite side of the work, particularly when drilling metal.

Note   When drilling metal, lubricate the bit to help cool the cutting edges and produce a smoother finished hole. A small amount of non-combustible cutting oil makes a good lubricant for drilling softer metals. Wood drilling requires no lubrication.

When drilling deep holes, periodically pull the drill bit partially out of the hole. This aids in clearing the hole of shavings and excess materials.

CAUTION   When drilling into or through a wall, be aware of what you might hit. Spaces between studs often contain electric wiring, plumbing, or insulation.

2.1.3 Safety And Maintenance Considerations

In addition to general safety rules and guidelines you learned in the Safety Module, there are some specific cautions to be aware of when working with drills. They are:

WARNING!   Do not use electric drills around **combustible** materials. Motors and/or bits can create sparks which can cause an explosion!

---

1. Always wear appropriate personal protective equipment.
2. Before connecting to the power source, make sure the trigger is NOT located in the "ON" position. It should be OFF. Always disconnect the power source before changing bits or maintenance.
3. Be sure to use the right bit for the job.
4. Make sure the drill bit is securely tightened in the chuck before starting the drill.
5. Make sure the chuck key is removed from the chuck before starting.
6. Do not use the trigger lock.
7. Always use a sharp bit.
8. Never ram the drill while drilling. This chips the cutting edge and damages the bearings.
9. Hold the drill with both hands, applying steady pressure. Let the drill do the work.
10. Be sure to determine in advance what is inside the wall or on the other side of the work material when doing work that involves cutting through a wall or partition. Avoid hitting water lines or electrical wiring.
11. *Always* use safety glasses.

12. Drills require little maintenance. Many have gears and bearings which are lubricated for life. Some drills have a small hole in the case to lubricate the motor bearings. Apply about three drops of oil occasionally, but don't overdo it. Excess lubricant may leak onto electrical contacts and burn the copper surfaces.

13. Keep the drill's air vent clean with a small brush or small stick. Ventilation is crucial to the maintenance and safety of a drill.

14. Attach the chuck key to the power cord when not in use. This prevents key loss.

15. DO NOT operate electric power tools without proper **ground fault protection**.

2.1.4 Performance/Laboratory Exercise

Supplies needed:

> Power drill
> Scrap wood pieces
> Scrap metal pieces
> C-clamps or bench vise
> Appropriate Personal Protective Equipment

Try this:
1. Review all safety and "How To" steps.
2. Load an appropriate bit into the chuck for drilling wood. Tighten and remove the key.
3. Stand the wood on a workbench or tabletop, securing it with one hand, C-clamp, or vise. Drill several holes through the wood.
4. Repeat for metal (remember: switch bits!).

Check to be sure that:
1. You are wearing appropriate personal protective equipment.
2. The chuck is well tightened.
3. The chuck key is removed before drilling.
4. You use a punch to start the hole.
5. The drill is perpendicular to the surface being drilled.
6. Relax the pressure when the bit is about to emerge from the opposite side of the work.

2.2.0 CORDLESS DRILL

Cordless power drills are especially useful for working in awkward areas or in areas where a power source is difficult to come by. Because of this flexibility, they are very popular tools.

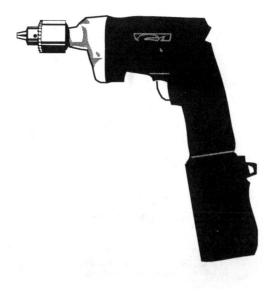

Figure 4. Cordless Drill

2.2.1 Description

Cordless drills *(Figure 4)* usually contain a rechargeable battery pack that runs the motor. The pack is detachable and can be plugged into a battery charger. Some chargers can recharge the battery pack in an hour, while others require more time. Workers who use cordless drills extensively usually carry an extra battery pack with them. Some cordless drills have adjustable clutches that allow the drill motor to also serve as a power screwdriver.

2.2.2 How To Use A Cordless Drill

Use the cordless drill the same way you use a power drill.

2.2.3 Safety And Maintenance Considerations

Follow the same safety practices used for a power drill.

2.3.0 HAMMER DRILL

The hammer drill, named so because of its hammering-type pounding action, lets you drill into concrete, brick, or tile. It rotates and hammers at the same time and drills much faster than regular drills. *(See Figure 5.)*

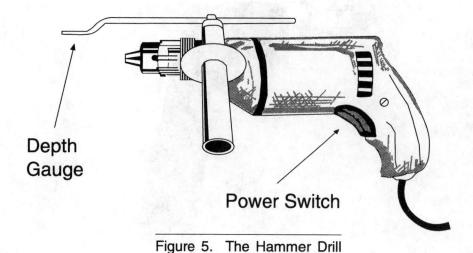

Depth
Gauge

Power Switch

Figure 5. The Hammer Drill

2.3.1 Description

The depth gauge is used to measure the depth of the hole being drilled and can be set to the depth of the hole needed.

Special hammer drill bits are required because they are designed to take the higher impact of the heavier materials. Percussion and masonry bits *(Figure 6)* are used with some hammer drills.

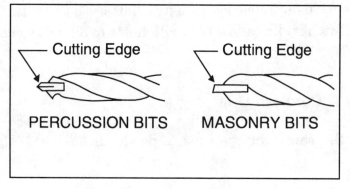

Cutting Edge   Cutting Edge

PERCUSSION BITS   MASONRY BITS

Figure 6. Hammer Drill Bits

2.3.2 How To Use A Hammer Drill

Step 1   Wear appropriate personal protective equipment.

Step 2   Follow the same procedures for using a power drill.

Step 3   Most hammer drills will not hammer until you put pressure on the drill bit.

You can adjust the drill's blows per minute by turning the adjustable ring. *(See Figure 7.)*

Step 4   The hammer action stops when you stop applying pressure to the drill.

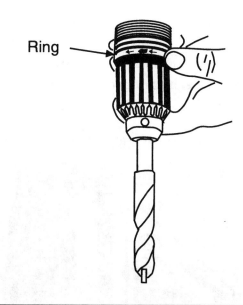

Ring

Figure 7. Adjustable Ring on a Hammer Drill

2.3.3 Safety And Maintenance Considerations

Follow the same safety practices used for a power drill.

2.3.4 Performance/Laboratory Exercise

Supplies needed:
　　Appropriate personal protective equipment
　　Hammer drill
　　Masonry bit
　　Concrete block

Try this:
1. Review all safety and "How To" steps.
2. Load an appropriate bit for drilling masonry into the chuck. Tighten and remove the key.
3. Use the hammer drill to drill a hole in the concrete block.

Check to be sure that:
1. You are wearing appropriate personal protective equipment.
2. The chuck is well tightened.
3. The chuck key is removed before drilling.
4. You use a punch to start the hole.
5. The drill is perpendicular to the surface being drilled.
6. Relax the pressure when the bit is about to emerge from the opposite side of the work.

2.4.0   ELECTRO-MAGNETIC DRILL

The electro-magnetic drill *(Figure 8)* is a portable drill mounted on an electro-magnetic base. It is used for drilling thicker metal. When placed on metal and energized (powered on), the magnetic base will hold the drill in place for drilling. The drill can also be rotated on the base.

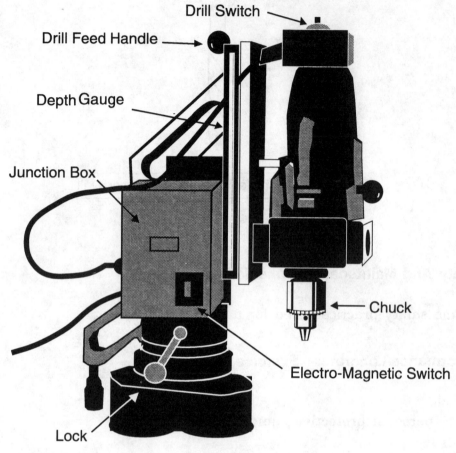

Figure 8.   Electro-Magnetic Drill

2.4.1   Description

The electro-magnetic base is controlled by a switch on the junction box. When the switch is turned on, the magnet holds the drill in place on a **ferromagnetic** metal surface. The switch on the top of the drill turns the drill on and off. The depth gauge is used to measure the depth of the hole being drilled and can be set to the depth of the hole needed.

2.4.2   How To Set Up An Electro-Magnetic Drill

Note　　　The *use* of this tool is addressed in the specific craft areas that use it. For now, we will describe the *set up* procedures for the drill only.

Step 1   Wear appropriate personal protective equipment.

Step 2   Place the drill face down into the metal holder.

Step 3   Power ON the electro-magnetic switch (NOT the drill). This holds the drill in place.

Step 4   Unlock the base-lock and rotate the drill on its base if necessary.

Step 5   Lock the drill in place.

Step 6   Set the depth gauge to the depth of hole needed.

Step 7   Position material securely on the drilling surface with C-clamps.

2.4.3   Safety And Maintenance Considerations

1.   Make sure that the material is securely clamped before proceeding. Unsecured material can become a lethal (deadly) projectile (flying object)!

2.   Securely support the drill before turning the electro-magnetic drill off! It will fall to the surface if you do not hold it while turning off the power.

2.5.0   PNEUMATIC DRILL (AIR HAMMER)

Pneumatic drills *(Figure 9)* are powered by compressed air from an air hose. They have many of the same parts, controls, and uses as an electric drill. They are the drill of choice when there is no electric source in the vicinity.

WARNING!   When working in areas containing combustible materials,
                    be sure to use a non-sparking drill.

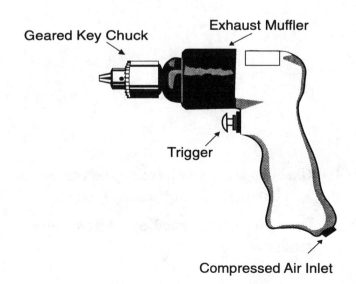

Figure 9. The Pneumatic Drill

2.5.1 Description

Common sizes of drills are 1/4", 3/8", and 1/2". The size indicates the maximum shank **diameter** that can be gripped in the chuck - not the drilling capacity.

2.5.2 How To Use A Pneumatic Drill

Step 1   Wear appropriate personal protective equipment.

Step 2   Make sure that the air pressure is shut off at the main air outlet.

Step 3   Hold the coupler at the end of the air supply line, slide the ring back, and slip the coupler on the connector or nipple that is attached to the air drill. *(See Figure 10.)*

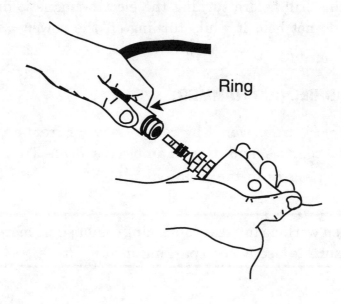

Figure 10. Joining the Coupler with the Connector

Step 4   Check to see if you have a good connection. A good coupling cannot be taken apart without first sliding the ring back.

Step 5   Once you have a good connection, turn the air supply valve on. The drill is now ready to use.

Step 6   When work is completed:

     A. Shut off the air supply to the hose. Use the trigger to "bleed off" excess pressure.

     B. Vent the compressed air in the hose by squeezing the trigger.

     C. Disconnect the drill from the hose.

2.5.3   Safety And Maintenance Considerations

Follow the same safety practices used for a power drill.

2.5.4   Performance/Laboratory Exercise

Supplies needed:

Personal Protective Equipment	Air hose
Pneumatic drill	Air source
Scrap wood pieces	Bits
Scrap metal pieces	Coupler
C-clamps or bench vise	

Try this (set up only):
1. Review all safety and "How To" steps.
2. Connect the coupler. Check for good connection.
3. Load an appropriate bit for drilling wood into the chuck. Tighten and remove the key.

Check to be sure that:
1. You are wearing appropriate personal protective equipment.
2. Air pressure is off at main outlet.
3. Coupler is connected securely.
4. When done, turn off air supply and bleed off excess pressure.
5. Disconnect drill from hose.

2.6.0   SELF-CHECK REVIEW / PRACTICE QUESTIONS 2

1. The most common use of the power drill is to:

    A. Cut down trees
    B. Hammer nails
    C. Make holes by driving drill bits into wood, metal, plastic, etc.
    D. Carve letters

2. What does the trigger switch on the power drill do?

    A. Controls the power. The harder you pull, the faster the speed.
    B. Shoots nails into walls.
    C. Blows air to clean the work area.

3. Should the point of the drill bit be sharper for soft materials or harder ones?

    A. Soft materials
    B. Harder materials

4. Use a chuck key to secure the bit into the drill chuck.

    A. True
    B. False

5. You should leave the chuck key in the chuck when operating the drill.

    A. True
    B. False

6. If the drill bit gets stuck in the material while drilling, what should you do *first*?

    A. Unplug the drill.
    B. Release the trigger switch.
    C. Back the drill out of the material.
    D. Switch the direction of the reversing switch, then back out the drill.

7. With a hammer drill, you can drill into:

    A. Dirt
    B. Concrete, brick, or tile
    C. Cars

8. The electro-magnetic drill is a:

   A. Drill that lights up in the dark
   B. Drill that locates metallic objects in the ground
   C. Portable drill mounted on an electro-magnetic base
   D. Portable drill mounted on a ceramic base

9. The pneumatic drill is powered by:

   A. Gas
   B. Water
   C. Air
   D. Gunpowder

10. Cordless drills are powered by:

   A. Microwave
   B. Rechargeable battery pack
   C. Infrared
   D. Solar energy

11. Identify each of these drills *(Figure 11)* and place their corresponding number next to their names on the following list:

   A._____Pneumatic Drill
   B._____Electro-Magnetic Drill
   C._____Cordless Drill
   D._____Hammer Drill

1.

2.

Figure 11. Identify the Drills

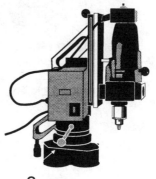
3.

4.

3.0.0 SAWS

Using the right type of saw for the job will make your work much easier. Always make sure that the blade is compatible with the material being cut out. The various types of power saws we will be addressing in this section are the:

- Circular Saw
- Saber Saw
- Reciprocating Saw
- Portable, Hand-held Band Saw
- Hand Jig Saw

3.1.0 CIRCULAR SAW

Many years ago a manufacturing company named "Skil" made power tool history by introducing the portable circular saw. Today there are dozens of models available made by many different manufacturers, yet it is not unusual to hear a portable circular saw referred to as a "Skil-saw." Other names you might hear are "cutoff saw," "utility saw," and "builder's saw." *(See Figure 12.)*

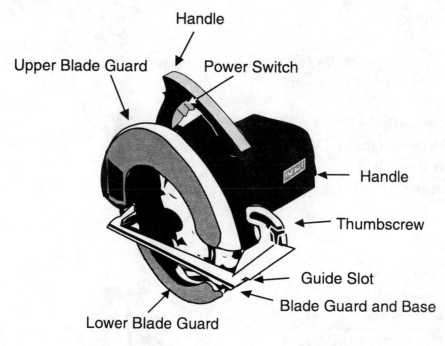

Figure 12. The Circular Saw

3.1.1 Description

Saw size is indicated by the diameter of the circular blade. Saw blade diameters are from 4-1/2" to 12", with the 7" to 8" size being the most popular. Blade speed is stated in r.p.m.s when the blade is running free.

The handle of the circular saw has a trigger switch that starts the saw. The teeth of the blade point in the direction of the rotation. The lower blade guard rotates to allow the saw to cut and also protects the blade from damage while it is rotating. The blade guard also protects you from the blade and from flying debris.

CAUTION   NEVER use the saw without the lower blade guard.

3.1.2   How To Use A Circular Saw

Step 1   Wear appropriate personal protective equipment.

Step 2   Properly secure the material to be cut. If the work isn't heavy enough to sit on its own without moving about, weight or clamp it down.

Step 3   Make your cut mark with a pencil or other marking tool.

Step 4   Place the front edge of the base plate on the work so the guide notch is in line with the cut mark.

Step 5   Grip the saw with two hands, as shown below in *Figure 13*.

Figure 13.   When You Can, Use a Two-Hand Grip

Step 6   Start the saw. *After* the blade has revved up to full speed, move the tool forward to start the actual cutting.

Note that the saw **kerf** has width, so the actual cut must be made on the waste side of the material.

Step 7   As you approach the end of the cut, the guide notch area of the baseplate will be off the work. Use the blade as your guide.

Step 8   Release the trigger switch. The blade stops rotating.

3.1.3 Safety And Maintenance Considerations

1. Wear appropriate personal protective equipment.
2. Check to see that the blade is tight.
3. Check to see that the retractable guard is functioning properly before connecting the saw to the power source.
4. Never reach underneath the work while operating the saw.
5. Never stand directly behind the work. Always stand to one side.
6. Do not attempt to saw small pieces while holding them in the hand. Use a clamp instead.
7. Always keep blades clean and sharp to reduce friction and possible kickback. Blades can be cleaned with hot water or mineral spirits.
8. Never force the saw through the work. This causes binding and overheating and possible injury to the operator.
9. Whenever possible, keep both hands on the saw while operating it.
10. The most important maintenance performed on a circular saw is at the lower blade guard. Sawdust accumulates in the retracting mechanism and causes the guard to stick. Unless the guard moves quickly over the blade after making a cut, damage may be caused when setting the saw down while the blade is still coasting and exposed. Remove sawdust from blade guard area. Remember, always disconnect the power source before maintenance.
11. To avoid the dangerous consequences of possible injury to yourself or your materials, check frequently to see that the guard snaps shut quickly and smoothly. Rubbing or binding can be eliminated by first disconnecting the power, removing the blade, and then cleaning the interior with alcohol or mineral spirit.
12. Avoid lubricating the guard with oil or grease because this could cause sawdust to stick in the mechanism.
13. Be aware of where the power cord is. Many have been cut by accident!
14. Be sure to determine in advance what is inside the wall or on the other side of the work material when doing work that involves cutting through a wall or partition. Avoid hitting water lines or electrical wiring.

3.1.4 Performance/Laboratory Exercise

Supplies needed:
    Appropriate personal protective equipment
    Circular saw
    Large wood scraps
    Bench vise or C-clamps
    Safety glasses

Try this:
Cut three pieces of wood of varying lengths (instructor will assign lengths).

Check to be sure that:
1. You are wearing appropriate personal protective equipment.
2. Check blade and guide notch.
3. Secure the material to be cut.
4. Mark the material to be cut.
5. Use the base plate properly.
6. Grip with two hands (as much as possible).
7. Let blade rev up before cutting.

3.2.0   SABER SAW

The saber saw *(Figure 14)* is one of the most versatile of portable power tools. It can make straight or curved cuts. It can also make its own starting hole when a cut must begin in the middle of a board or panel. Many models are available with tilting shoe plates to permit **bevel cutting**. Most models can be adapted in seconds to cut metal or plastics, as well as wood, simply by changing blades.

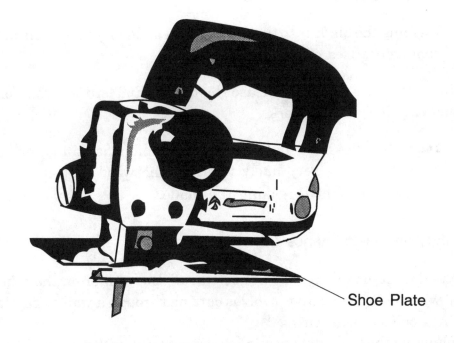

Shoe Plate

Figure 14. The Saber Saw

3.2.1   Description

Interchangeable blades enable the saber saw to cut a wide variety of materials, ranging from wood and metal to wall boards and ceramic tiles.

Saber saws come with various speed options including single speed, two speed, and variable speed.

The variable-speed saber saw is capable of cutting at low and high speeds. The low-speed setting is best for cutting hard materials and the high-speed for soft materials.

An important part of the saber saw is the shoe or footplate. Its broad surface helps to keep the blade aligned. It prevents the work from vibrating and allows the blade teeth to bite into the material.

3.2.2 How To Use A Saber Saw

Step 1   Wear appropriate personal protective equipment.

Step 2   If possible, to avoid vibration, clamp the work to a sturdy pair of saw horses or a vise.

Step 3   When starting to cut from the edge of a board or panel, be sure the front of the saw's shoe plate is resting firmly on the surface of the work *before* the saw is started. The blade *should not* be in contact with the work at this stage.

Step 4   Start the saw (pull the trigger) and move the blade gently into the work. Do not push the blade into the work.

Note   Pushing the blade into the work with a sudden bump is one of the most common causes of blade breakage.

Step 5   When the cut is finished, release the trigger and let the blade come to a stop before removing it from the work.

Note   Do not lift the blade out of the work while it is still running. If you do so, the tip is very likely to strike the wood surface, marring the work and possibly breaking the blade.

3.2.3 Safety and Maintenance Considerations

1. Be sure to determine in advance what is inside the wall or on the other side of the work material when doing work that involves cutting through a wall or partition. Avoid hitting water lines or electrical wiring.
2. Before plugging the tool into a power source, make sure the switch is in the OFF position.
3. Always use a sharp blade and never force the tool through the work.
4. Secure the material you are working with to reduce vibration and ensure safe working conditions.
5. When installing a blade in the collar of the plunger, make sure it is inserted as far as it will go and tighten the set screw securely. Always disconnect the power source before changing blades or performing maintenance.
6. When replacing a broken blade, replace any shattered pieces that may be lodged inside the collar.

7. Do not force or lean into the blade. This could result in you pitching forward or your hands going into the work surface.
8. Always wear appropriate personal protective equipment around power tools.

3.2.4 Performance/Laboratory Exercise

Supplies needed:
    Appropriate personal protective equipment
    Saber saw
    Scrap wood
    Bench vise or C-clamps

Try this:
Practice making cuts in the scrap wood with the saber saw.

Check to be sure that:
1. You are wearing appropriate personal protective equipment.
2. Check blade and guide notch.
3. Secure the material to be cut.
4. Mark the material to be cut.
5. Use the base plate properly.
6. Grip with two hands (as much as possible).
7. Let blade rev up before cutting.
8. Do not push saw through work.
9. Let blade stop running before lifting from work.

3.3.0 RECIPROCATING SAW

Both the saber saw and the reciprocating saw are used to make straight and curved cuts. They often are used to cut irregular shapes and holes in plaster, plasterboard, plywood, studs, metal, or most other materials that can be cut with a saw.

Both have straight blades that move back and forth and are guided in the direction of the cut. But here's the difference: the saber saw's blade moves up and down. The reciprocating saw's blade - and note that **reciprocate** means "to move alternately, back and forth" - moves, not surprisingly, back and forth. This saw is a more heavy-duty saw than the saber saw. It can drive longer and tougher blades than those you can place in a saber saw. Also, because of its design, you can get into more inaccessible places with it.

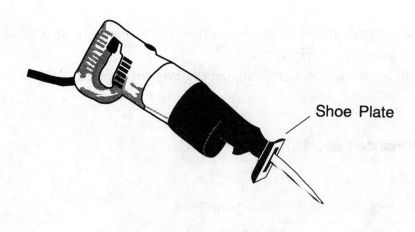

Shoe Plate

Figure 15. The Reciprocating Saw

There are many times where a saber saw can be efficiently controlled with one hand. It will be a rare exception when you might be able to control a reciprocating saw with one hand. Usually, you will need to use both hands gripping it firmly. *(Figure 15.)*

3.3.1 Description

Like the saber saw, reciprocating saws come with various speed options including single speed, two speed, and variable speed.

The reciprocating saw is capable of cutting at low and high speeds. The low-speed setting is best for metal work and the higher speed for sawing wood and other relatively soft materials.

The shoe, or foot plate, may have a swiveling action or it may be fixed. Whatever the design, it's there to provide a brace-point for the sawing operation.

3.3.2 How to Use a Reciprocating Saw

Step 1   Wear appropriate personal protective equipment.

Step 2   If possible, to avoid vibration, clamp the work to a sturdy pair of saw horses or a vise.

Step 3   Set the saw to the desired speed range (speed selection made at the trigger on-off switch). Remember:

    • lower speeds for metal work
    • higher speeds for sawing wood and other relatively soft materials.

Step 4   Grip the saw with both hands. Place the foot plate firmly against the workpiece. *(See Figure 16.)*

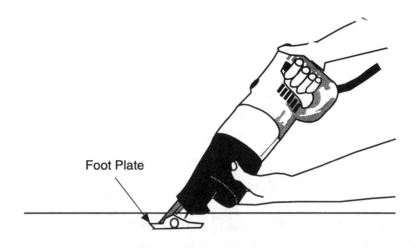

Foot Plate

Figure 16. Use Both Hands - Place Foot Against Workpiece

Step 5   Squeeze the trigger on-off switch. The blade reciprocates back and forth, cutting on the backstroke.

3.3.3   Safety and Maintenance Considerations

1. Always wear appropriate personal protective equipment around power tools.
2. Be sure to determine in advance what is inside the wall or on the other side of the work material when doing work that involves cutting through a wall or partition. Avoid hitting water lines or electrical wiring.
3. Always disconnect the power source before changing blades or performing maintenance.

3.3.4   Performance/Laboratory Exercise

Supplies needed:
    Appropriate personal protective equipment.
    Reciprocating saw
    Scrap wood
    Bench vise or C-clamps
    Safety glasses
Try this:
Practice making cuts in the scrap wood with the reciprocating saw.

Check to be sure that:
1. You are wearing appropriate personal protective equipment.
2. Set to correct speed for material being cut.
3. Secure the material to be cut.
4. Mark the material to be cut.

5. Use the base plate properly.
6. Grip with two hands (as much as possible).
7. Let blade rev up before cutting.
8. Do not push saw through work.
9. Let blade stop running before lifting from work

3.4.0   PORTABLE HAND-HELD BAND SAW

The portable hand-held band saw *(Figure 17)* is used where it is more advantageous to move the saw to the work rather than the work to the saw. It can be used to cut pipe, metal, plastics, wood, and irregularly shaped materials. It is especially good for cutting heavy metal - but it will also do fine curved scroll cutting.

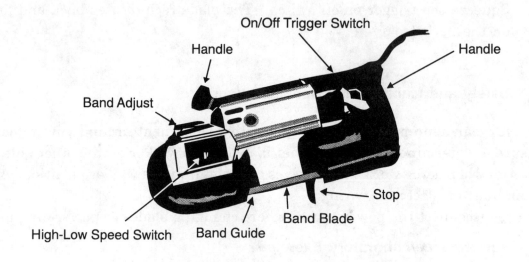

Figure 17. The Portable Hand-Held Band Saw

3.4.1   Description

The band saw has a continuous one-piece band blade that runs in one direction around guides located at either end of the saw. The blade is a thin, flat piece of steel. It is sized according to the diameter of the revolving pulleys that drive and support the blade. The saw often has a choice of speeds.

3.4.2 How To Use A Portable Band Saw

Step 1   Wear appropriate personal protective equipment.

Step 2   Place the stop located below the band firmly against the object to be cut. This prevents the saw from bouncing against the object and breaking the band.

Step 3   Begin gently pulling the trigger.

Only the slightest pressure is needed to make a good clean cut since the weight of the saw itself is sufficient to produce adequate leverage for cutting.

Note   The tool cuts on the pull, not the push.

Step 4   The band saw cuts best at a low speed. High speed causes the teeth to rub and not cut. Rubbing also causes heat, which will cause the blade to wear out quickly.

3.4.3 Safety and Maintenance Considerations

1. Always wear appropriate personal protective equipment.
2. The blade of a portable band saw binds very easily. Never force a portable band saw. Let the saw do the cutting.
3. The blades should be waxed periodically with a special lubricant. Always disconnect the power source before maintenance.
4. Be sure to determine in advance what is inside the wall or on the other side of the work material when doing work that involves cutting through a wall or partition. Avoid hitting water lines or electrical wiring.
5. Only use a band saw that has a stop.

3.4.4 Performance/Laboratory Exercise

Supplies needed:
    Appropriate personal protective equipment
    Portable band saw
    Metal pipe
    Safety glasses
    Bench vise

Try this:
1. Place the pipe securely in the vise.
2. Cut completely through the pipe with the bandsaw.

Check to be sure that:

1. You are wearing appropriate personal protective equipment.
2. The stop is placed firmly against the object to be cut.
3. Secure the material to be cut.
4. Mark the material to be cut.
5. Grip with two hands (as much as possible).
6. Gently pull trigger. Let blade rev up before cutting.
7. Pull the tool to cut.
8. Let blade stop running before lifting from work.

3.5.0   PORTABLE HAND JIG SAW

Jig saws *(Figure 18)* have extremely fine blades which makes them great tools for doing delicate and intricate work, such as cutting out patterns or irregular shapes from wood or from thin soft metals. They are also one of the best tools for cutting circles.

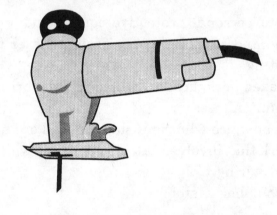

Figure 18.   The Portable Hand Jig Saw

3.5.1   Description

Most jig saws are capable of being operated at various blade speeds. Harder materials are best sawed at slower speeds. Though the tool is often thought of for use only in craft applications, the jigsaw can also work with fairly heavy blades and average maximum depth of cut turns to about 2". Thus you can work with some fairly heavy work. With the correct blade and speed, the jigsaw can be used to cut hard and soft metals and other non-wood materials such as leather, plastics, laminate, and paper. The blade of this saw moves up and down.

3.5.2 How To Use The Portable Jig Saw

Step 1 Wear appropriate personal protective equipment.

Step 2 Check blade to see that it is the right blade for the job and that it is sharp and undamaged.

Step 3 Plug saw into a power source, making sure that the saw is grounded.

Step 4 Measure and mark the work.

Step 5 Hold work securely on solid support and align the saw blade with the mark on the work.

Step 6 Hold base of saw firmly on work and check to be sure that blade clears work.

Step 7 Start saw, advance forward into work, and carefully guide saw blade into layout line.

Step 8 Continue feeding saw into work as fast as possible without forcing it. *(See Figure 19.)* Complete the cut.

Step 9 Turn off saw immediately after the cut is made.

Step 10 After blade has stopped moving, lay saw on its side. Unplug saw from power source.

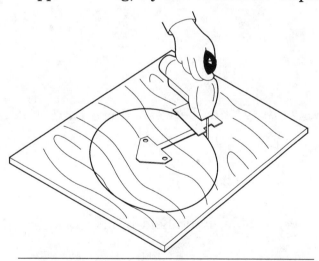

Figure 19. Cutting a Circle with the Jig Saw

3.5.3 Safety And Maintenance Considerations

1. Always wear appropriate personal protective equipment.
2. A very slow speed is essential for avoiding blade damage.
3. If you are using the jigsaw to cut metal, use a metal cutting blade. Lubricate the blade with something like beeswax to help make tight turns and reduce the possibility of breakage. Always disconnect the power source before maintenance.
4. If you find that sawdust is not blowing away from the cutting line, remove and clean the air tube.

3.6.0 SELF-CHECK REVIEW / PRACTICE QUESTIONS 3

1. The reason why portable circular saws are often referred to as "Skil" saws is:

   A. You need a lot of skill to operate one
   B. It's the name of the company that first introduced it

2. The size of the saw is indicated by the diameter of the blade.

   A. True
   B. False

3. Why can't you use a circular saw that doesn't have a lower blade guard?

   A. It's unattractive.
   B. The saw blade may fall off.
   C. It protects you from the blade and from flying particles.

4. Saber saws can make their own starting hole when a cut must begin in the middle of a board or panel.

   A. True
   B. False

5. Which is the more "heavy duty" saw between these two?

   A. Saber saw
   B. Reciprocating saw

6. A great advantage of the portable, hand-held band saw over other saws is that

   A. It can cut metal, plastics, wood
   B. It can be moved easily to the job site

7. The best tool for cutting delicate letters from wood would be a

   A. Cordless drill
   B. Saber saw
   C. Jack hammer
   D. Hand jig saw

8. Identify each of these saws in *Figure 20* below and place their corresponding number next to their names on the following list:

A. _____ Circular saw
B. _____ Saber saw
C. _____ Reciprocating saw
D. _____ Portable, hand-held band saw
E. _____ Hand-held jig saw

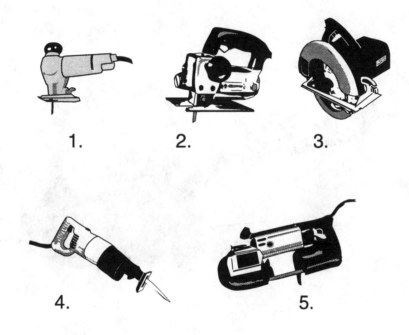

Figure 20. Identify these Saws

4.0.0   GRINDERS AND SANDERS

Grinding tools can drive an endless assortment of **abrasive** wheels, brushes, buffs, drums, bits, saws, and discs. These wheels are available in a variety of materials and **grits**. They can drill, cut, smooth, and polish. They can be used to shape or sand wood or metal. They can be used on plastics and will mark steel or glass. They can sharpen and engrave. As you can see, grinders are flexible tools!

The types of grinders and sanders addressed in this section are:

- Angle Grinder (side grinder)
- End Grinder
- Bench Grinder

4.1.0. ANGLE GRINDERS

The angle grinder, also referred to as a side grinder, is used to grind away hard, heavy materials and for surface grinding such as pipe, plates, or welds. *(See Figure 21.)*

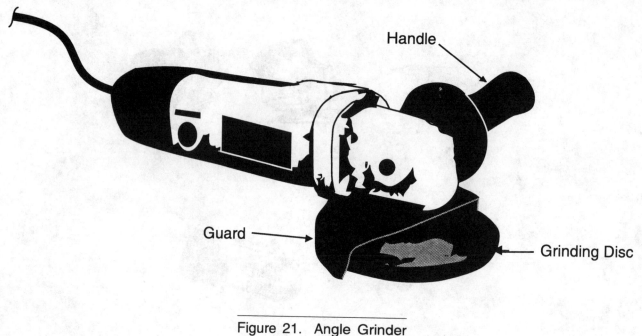

Figure 21. Angle Grinder

4.1.1 Description

The angle grinder has a rotating grinding disc positioned at a right angle to the motor shaft.

4.2.0 END GRINDERS

End grinders are also called horizontal grinders and/or pencil grinders. These smaller grinders are used to smooth the inside of materials such as pipe. *(See Figure 22.)*

Figure 22. End Grinder

4.2.1 Description

The grinding disc on the end grinder rotates in line with the motor shaft. Grinding is also done with the circumference (outside) of the grinding disc.

4.2.2 How To Use An Angle/End Grinder

Step 1   Wear appropriate personal protective equipment.

Step 2   Secure material in a vise or clamp it to a bench top.

Step 3   To use an **end grinder**, grip the grinder at the shaft end of the tool with one hand and "palm" the other end of the tool in your other hand.

            To use an **angle grinder**, use one hand on the handle of the grinder and one on the trigger.

Step 4   Finish the work by removing any loose materials with a wire brush.

4.2.3 Safety And Maintenance Considerations

1. Be sure to have firm footing and a firm grip before using grinders. They have a tendency to pull you off balance.
2. Never position a hand that is holding the work so that it is in line with the grinder.
3. For all grinders: Before starting the grinder, make sure the grinding disc is secured and is in good condition. Always disconnect the power source before maintenance.
4. Always use a spark deflector/shield as well as proper eye protection.
5. Make sure all guards are in place.
6. Adjust tool rest to 1/8" of the wheel.
7. Shut off power and do not leave tool until the grinding disc has come to a complete stop.
8. Never use an end grinder unless it is equipped with the guard that surrounds the grinding wheel.
9. Always wear appropriate personal protective equipment around power tools.
10. Direct sparks and debris away from personnel and any **hazardous materials**.
11. Use a flame-retardant blanket to catch falling sparks when grinding on an elevated location.

4.2.4 Performance / Laboratory Exercise

Supplies needed:
   Appropriate personal protective equipment, including proper eye protection (including a spark deflector)
   Angle grinder
   Rough-edged scrap metal
   Bench vise or C-clamps

Try this:
Grind away the excess rough edges of the rough-edged scrap metal.

Check to be sure that:
1. You are wearing appropriate personal protective equipment, including proper eye protection (including a spark deflector).
2. Material is secured in vise or C-clamp.
3. Make sure grinding disc is secured and in good condition.
4. The tool is properly held.
5. You have firm footing.
6. The guards are in place.
7. Tool rest is adjusted to 1/8" of the wheel.
8. Finish work by removing loose materials with wire brush.
9. Do not walk away from tool while disc is moving.

4.3.0 BENCH GRINDER

Bench grinders *(Figure 23)* are electrically powered, stationary grinding machines. They usually have two grinding wheels that are used for grinding, rust removal, and metal buffing. They are also great for renewing worn edges and maintaining the sharp edges of cutting tools. Remember learning about the danger of "mushroomed" cold chisel heads in the *Basic Hand Tools* module? They can be smoothed with the bench grinder!

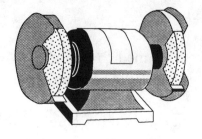

Figure 23. The Bench Grinder

4.3.1   Description

The heavy-duty grinder wheels range in size from 6-3/4" to 10" in diameter. Each wheel's maximum speed is given in r.p.m.s. Never use a grinding wheel above its rated maximum speed.

Other attachments for the bench grinder include knot wire brushes used for removing rust, scale, and file marks from metal surfaces. Cloth buffing wheels are also available for bench grinders. These are used for polishing and buffing metal surfaces.

4.3.2   How To Use A Bench Grinder

Step 1   Wear appropriate personal protective equipment.

Step 2   Always use the adjustable tool rest on the bench grinder. It serves as a support when grinding or bevelling metal pieces. There should always be a 1/8" gap between the tool rest and the wheel.

CAUTION   Never change the adjustment of these tool rests when the grinder is on or when the grinding wheels are spinning.

Step 3   Let the wheel come up to full speed before you apply the work.

Step 4   Keep the metal you are grinding cool by having water on hand to cool the item frequently. If the metal gets too hot, it can destroy the **temper** of the tool.

Step 5   Whenever possible, work on the face of the wheel. Working on the side of the wheel is necessary for many jobs, but inspect it frequently to be sure you do not reduce wheel thickness to the point where further use becomes hazardous.

4.3.3   Safety and Maintenance Considerations

1. Always wear appropriate personal protective equipment.
2. Keep hands away from the grinding wheels.
3. Never wear loose clothing when grinding. It can get caught in wheels.
4. When finished using the bench grinder, shut it off.
5. Always make sure the bench grinder is disconnected before changing grinding wheels.
6. Never use a grinding wheel above its rated maximum speed.
7. Let the wheel come up to full speed before you apply the work.
8. Always adjust the tool rests so they are within 1/8" of the wheel. This reduces the possibility of getting the work wedged between the rest and the wheel.

9. Test a wheel for cracks before you mount it. After doing a visual check for chipped edges and cracks, mount the wheel on a rod that you pass through the wheel hole. Tap the wheel gently on the side with a piece of wood. The wheel will ring clear if it is in good condition. A dull thud may indicate the presence of an unseen crack. Discard the wheel if this happens.

10. Metal grinding creates sparks, so keep the area around the grinder clean.

4.4.0   SELF-CHECK REVIEW / PRACTICE QUESTIONS 4

1. If you want to grind away hard, heavy materials, you can use an angle grinder.

   A. True
   B. False

2. The stationary grinding machine is the _____ .

   A. Angle grinder
   B. End grinder
   C. Bench grinder

3. On the bench grinder, there should always be a 1/8" gap between the adjustable tool rest and the wheel.

   A. True
   B. False

4. Identify each of these grinders and sanders (Figure 24) and place their corresponding number next to their names on the following list:

   A. _____ Angle grinder
   B. _____ End grinder
   C. _____ Bench grinder

1.

2.

3.

Figure 24. Identify These Grinders and Sanders

5.0.0   MISCELLANEOUS POWER TOOLS

This section addresses:

- Jack Hammer (paving hammer)
- Porta-Power (hydraulic jack)
- Powder-Actuated (explosive) Tools

5.1.0   JACK HAMMER (PAVING HAMMER)

A jack hammer is a type of pneumatic percussion hammer. It is used for large scale demolition work such as tearing down brick and concrete walls and breaking up concrete or pavement. These hammers do not rotate like hammer drills. They reciprocate. *(See Figure 25.)*

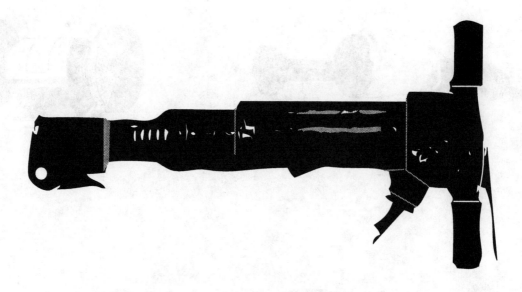

Figure 25. The Jack Hammer

5.1.1   Description

The jack hammer can weigh between 50 and 80 pounds, depending on its specific design and manufacturer.

On most jack hammers, a throttle is located on the T-handle. When the throttle is depressed, compressed air operates a piston inside the tool. The piston drives the steel cutting shank into the material to be broken. Attachments can be used for a variety of purposes and tasks (e.g., chisels, spades).

5.1.2   How To Set Up A Jack Hammer (Coupling Only)

Step 1   Wear appropriate personal protective equipment.

Step 2   Make sure that the air pressure is shut off at the main air outlet.

Step 3   Hold the coupler at the end of the air supply line, slide the ring back, and slip the coupler on the connector or nipple that is attached to the air drill.

Step 4   Check to see if you have a good connection. A good coupling cannot be taken apart without first sliding the ring back.

Step 5   Once you have a good connection, turn the air supply valve on. The drill is now ready to use.

Step 5   The air hose must be connected properly and securely. Some fittings require the use of safety wires to prevent them from coming apart.

5.1.3 Safety And Maintenance Considerations

1. Be aware of what is below the material you are about to break. Be aware of existing water, gas, electricity, sewer, or telephone lines. FIND OUT BEFORE BREAKING THE PAVEMENT!
2. Always wear appropriate personal protective equipment around power tools.
3. Since some of these tools make a lot of noise, you *must* wear ear protection (ear plugs).

5.2.0 PORTA-POWER (HYDRAULIC JACK)

The portable hydraulic jack, or Porta-Power, has many uses. It is used for pushing heavy machinery and other heavy objects, for pulling wheels, bearings, gears, cylinder liners, and for straightening or bending frames.

5.2.1 Description

Porta-Powers have two basic parts - the pump and the cylinder. The two parts are joined by a high-pressure hydraulic hose. There are many different types of pumps and cylinders that can be used in different combinations for many types of jobs.

The pump applies pressure to the hydraulic fluid. The cylinder, sometime called a ram, applies a lifting or pushing force. Available in many sizes, cylinders are rated by the weight (in tons) they can lift and the distance they can move the weight. This distance is called stroke and is measured in inches. Hydraulic cylinders can lift weights to over 500 tons. Strokes may be from 1 inch to over 48 inches. Different cylinder sizes and ratings are used for different jobs.

5.2.2 How To Use A Porta-Power

Step 1   Wear appropriate personal protective equipment.

Step 2   Place the jack beneath the object (e.g., a car) to be lifted. You may have to use a wedge under other equipment to begin the lift.

Step 3   Pump the handle down and then release it. This raises the cylinder.

Note   If the leverage of the handle is 6:1, a common ratio, then here is what happens: If you apply 100 pounds of force to the handle, you can lift 6 tons!

Step 4   To lower the jack, open the return passage by means of a thumbscrew. The weight of the piston and its load then pushes the fluid in the cylinder back into the reservoir.

5.2.3   Safety And Maintenance Considerations

1. Check fluid level in the pump before use.
2. Make sure the hydraulic hose is not twisted.
3. Do not use a "cheater bar" (extension) on the pump handle.
4. Clear the work area when making a lift.
5. When lifting, make sure the cylinder is placed on a hard surface.
6. Do not move the pump if the hose is under pressure.
7. Watch for leaks.
8. Always wear appropriate personal protective equipment around power tools.

5.3.0   POWDER-ACTUATED (EXPLOSIVE) TOOLS

Powder-actuated tools have been used in the construction industry for years. All powder-actuated tools use the energy from a gunpowder charge to drive a fastener into steel or concrete. The fastener may be a stud bolt, a nail-like fastener, or a fastener of special design. The charge may be a metal-shell cartridge similar to a gun cartridge or it may be a powder pellet.

5.3.1   Description

In some powder-actuated tools, the energy from the powder charge is applied directly to the fastener to drive it. In others, the energy from the powder charge drives a piston which, in turn, drives the fastener with a hammer-like blow. Some powder-actuated tools have the capacity to drive 1/3" studs up to 3" long into steel or concrete in a matter of seconds.

5.3.2   How To Use Powder-Actuated Tools

Because of the potential danger that the misuse of these tools may cause, every operator is required to have an Accredited Operator's card issued by a qualified, AUTHORIZED instructor. This card has a serial number that is registered with the employer, the tool distributor, and the tool manufacturer. If you do not have this authorization, you may not use explosive power tools.

5.4.0 SELF-CHECK REVIEW / PRACTICE QUESTIONS 5

1. Jack hammers are used primarily to:

   A. Install telephone jacks
   B. Hammer nails
   C. Break up concrete or pavement
   D. Remove nails

2. One of the Porta-Power's uses is to push heavy machinery and other heavy objects.

   A. True
   B. False

3. The two basic parts of a Porta-Power are the:

   A. Cylinder and hose
   B. Reservoir and relief pump
   C. Pump and cylinder
   D. Hose and pump

4. The job of the pump on the Porta-Power is to apply pressure to the hydraulic fluid.

   A. True
   B. False

5. Porta-Power cylinders are rated by how much weight they can lift and:

   A. How much they weigh
   B. The amount of electro-magnetic material they have
   C. Their torque
   D. The distance they can move the weight

6. When lifting with a Porta-Power, you should make sure that the cylinder is placed on a hard surface.

   A. True
   B. False

7. The operator of an explosive powder actuated tool must have in his/her possession an Accredited Operator's card.

A. True
B. False

PERFORMANCE / LABORATORY EXERCISE

A. POWER DRILL

Supplies:

For this exercise, you will need the following: Power drill, Several bits (one appropriate for metal), Metal scraps, C-Clamp or vise.

Steps:

1. Choose appropriate bit for metal.

2. Load bit and remove chuck key.

3. Secure work.

4. Drill 2 holes.

B. POWER DRILL

Supplies:

For this exercise, you will need the following: Power drill, Several bits (one appropriate for wood), Wood scraps, C-Clamp or vise.

Steps:

1. Choose appropriate bit for wood.

2. Load bit and remove chuck key.

3. Secure work.

4. Drill 2 holes.

C. CIRCULAR SAW

Supplies:

For this exercise, you will need the following: Circular saw, Wood scraps, C-Clamp or vise, Combination square

Steps:

1. Check tightness of blade and operation of guard.

2. Mark wood and secure.

3. Cut the wood.

APPENDIX A: ANSWERS TO SELF-CHECK REVIEW / PRACTICE QUESTIONS

SELF-CHECK REVIEW 1

1. C
2. B
3. A
4. D
5. D
6. B
7. A

SELF-CHECK REVIEW 2

1. C
2. A
3. A
4. A
5. B
6. B
7. B
8. C
9. C
10. B
11. A:4
    B:3
    C:1
    D:2

SELF-CHECK REVIEW 3

1. B
2. A
3. C
4. A
5. B
6. B
7. D
8. A:3
    B:2
    C:4
    D:5
    E:1

SELF-CHECK REVIEW 4

1. A
2. C
3. A
4. A:3
   B:1
   C 2

SELF-CHECK REVIEW 5

1. C
2. A
3. C
4. A
5. D
6. A
7. A

notes

WHEELS OF LEARNING USER UPDATES

The NCCER makes every effort to keep these manuals up-to-date and free of technical errors. We appreciate your help in this process. If you have an idea for improving this manual, or if you find an error, a typographical mistake, or an inaccuracy in the *Wheels of Learning*, please write us, using this form or a photocopy. Be sure to include the exact module number, page number, a description of the problem, and the correction, if possible. We'll do our best to correct it in later editions. Thank you for your assistance.

Write: *Wheels of Learning*
National Center for Construction Education and Research
P.O. Box 141104
Gainesville, FL 32614-1104
Fax: 352-334-0932

WHEELS OF LEARNING USER UPDATE

Please let us know if you have found an inaccuracy, error, or other problem in a *Wheels of Learning* manual. Use this form or write us a letter. Please be sure to tell us the exact module name and module number, the page number, and the problem. Thanks for your help.

Craft _____ Module Name _____

Module Number _____ Page Number(s) _____

Description of Problem _____

_____

_____

(Optional) Correction of Problem _____

_____

_____

(Optional) Your Name and Address _____

_____

Introduction to Blueprints

Module 00105

Core Curricula Trainee Task Module 00105

INTRODUCTION TO BLUEPRINTS

Objectives

Upon completion of this module, the trainee will be able to:

1. Identify and recognize basic blueprint terms and symbols.
2. Relate information on prints to real parts and locations.

Prerequisites

None.

How To Use This Manual

During the course of completing this module, you will be taught and will practice identifying and recognizing commonly used blueprint terms and symbols. *Self-Check Review / Practice Questions* will follow the introduction of most terms/symbols. The answers to these written exercises are found in Appendix A, *Answers to Self-Check Review / Practice Questions*.

New terms will be introduced in **bold** print. The definition of these terms can be found in the front of this manual, under *Trade Terms Introduced In This Module*.

You will be referring to two different types of visuals in this document – figures and foldouts. Figures are found within the text itself. The foldouts are separate charts (see note below).

Required Student Materials

① Student Manual
② Set of full-size, sample blueprints
③ Pencils

C.S.I → CONSTRUCTION STANDARD INSTITUTE

IMPORTANT NOTE REGARDING BLUEPRINTS

A set of 24 blueprints is included with the *Wheels of Learning* Core Curricula, Trainee Module 00105, Introduction to Blueprints. Trainees should be sure to keep this set of prints because they are used for instruction in other *Wheels of Learning* training modules. The set is made up of ten 11" by 17" prints, and fourteen full size prints.

Course Map Information

This course map shows all of the *Wheels of Learning* task modules in the Core Curricula. The suggested training order begins at the bottom and proceeds up. Skill levels increase as a trainee advances on the course map. The training order may be adjusted by the local Training Program Sponsor.

Course Map: Core Curricula, Introduction To Blueprints

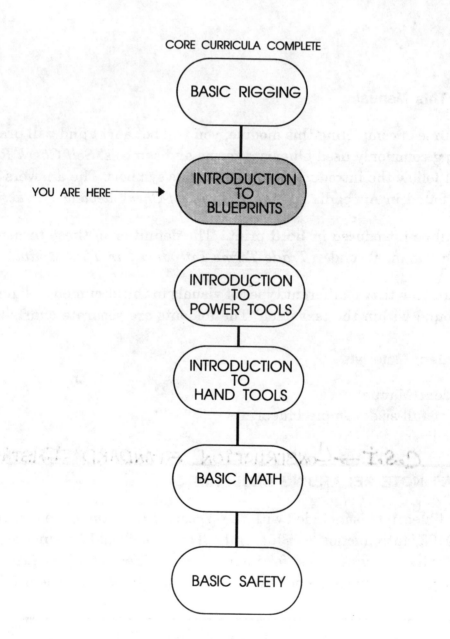

CORE CURRICULA COMPLETE

BASIC RIGGING

YOU ARE HERE → INTRODUCTION TO BLUEPRINTS

INTRODUCTION TO POWER TOOLS

INTRODUCTION TO HAND TOOLS

BASIC MATH

BASIC SAFETY

TABLE OF CONTENTS

Trade Terms Introduced in This Module:

Architect: One who designs and supervises the construction of buildings.

Auxiliary Drawings: Drawings that are supplementary to the working drawings. Examples are: Electrical Plan, HVAC Plan, Plumbing Plan, Door/Window Schedules, Foundation and Framing Plans.

Bay: A compartment set off from other compartments making up a given structure.

Blueprint: A photographic reproduction, as of architectural plans, shown as white lines on a blue background (also referred to as "working drawings" and "construction drawings").

Construction Drawings: Drawings used in actual construction of a building or part of a building. Also referred to as "Blueprints" and "Working Drawings."

Contour Lines: Show the rise and fall of land elevation on a site.

Cornice: A horizontal molded projection that crowns or completes a building or wall.

Detail Drawings: Enlarged views taken from a drawing. Used to show an area more clearly.

Dimensions: A definite measurement shown on a drawing as length, width, or height.

Drawing Title: Located within the title block, identifies the project.

Engineer: One who applies scientific principles for practical purposes in the design, construction and application of efficient and economical structures, equipment, and systems.

Elevation (ELEV): Height above sea-level (or other defined surface), usually expressed in feet. Elevation (ELEV) is given at each corner of the site.

Elevation Drawings: The side view of a building or object. Shows height and width but not depth.

HVAC: Heating, Ventilation, Air Conditioning

Legend: Explains or defines a symbol or special mark placed on the drawing.

North Arrow: Positions the drawings towards North (compass orientation)

Piers: Any of various vertical supporting structures.

Piping Symbols: Symbols for piping used to contain gas, oil, steam heat, and conduit wiring.

Plan: The top view of a building or object from about four feet above the level shown.

Scale: The ratio between the size of an object being drawn and its actual size. Also, a measuring device used for laying off distances.

Sectional Drawings: A cutaway drawing showing the inside of an object or building.

Specification Sheet: Also called **specs**. The written (usually typed) requirements issued by architects or engineers to establish general conditions, standards, and detailed instructions which are used with the blueprints.

Structural: Something made up of parts that are put together in a particular way.

Topographic: The physical features of an area.

Working Drawings: Drawings used in actual construction of a building or part of a building. Also referred to as "Blueprints" and "Construction Drawings."

1.0.0 INTRODUCTION

To begin with, let's answer the question: "What is a **blueprint**?"

A blueprint is a copy of a drawing in which, historically, the lines are white and the background is *blue* - thus, *blue*print. Today, however, the most common types of prints are blue or black lines on a white background. *(See Figure 1.)*

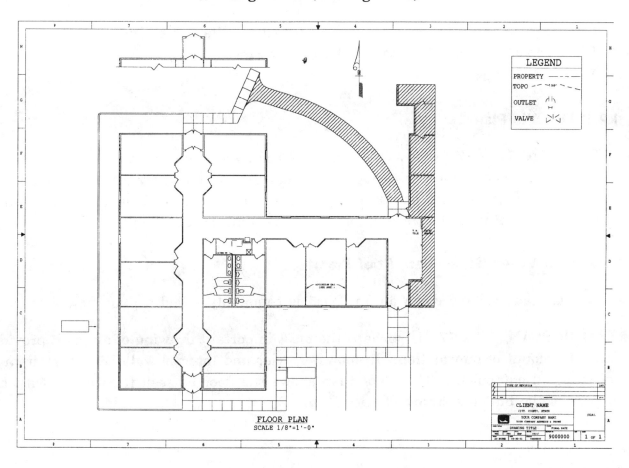

FLOOR PLAN
SCALE 1/8"=1'-0"

Figure 1. Blueprint of a Church Reconstruction (Floor Plan)

Construction blueprints are a means for the players on a construction job to communicate ideas and relate specific directions. These "players" are the **architects** and **engineers** who design and oversee the project; the bankers and owner who finance it; and the workers (you!) who construct it. Blueprints, together with the set of **specification sheets (specs)**, detail what is to be built, what materials are to be used, and how the job will be done.

1.1.0 SET OF BLUEPRINTS

The set of blueprints (also called "working drawings") form the basis of agreement and understanding that a building will be built as it is planned. Therefore, all persons who are involved in the planning, supplying and/or building of any structure should be able to read construction blueprints. A set of blueprints typically consists of the following: **working drawings** and **auxiliary drawings**.

1.2.0 WORKING DRAWINGS

Working Drawings consist of the following plans and drawings:

- Site Plan
- Elevation Drawings
- Detail Drawings
- Plan Views (Floor Plan, Roof Plan)
- Sectional Drawings

FOUR BASIC VIEWS OF A DRAWING

1.2.1 The Site Plan

The SITE PLAN *(11" x 17")* shows the location of the building on the site from a bird's-eye view. The **plan** also may include **topographic** features, such as **contour lines** and trees, and construction features such as walks, driveways and utilities. This is where it all starts - if the site is not acceptable, there is no reason to continue building!

1.2.2 Plan Views (Floor Plan, Roof Plan)

Any drawing made looking down on an object is commonly called a plan view.

The FLOOR PLAN *(11" x 17")* is perhaps the most important drawing of all, as it provides the largest amount of information. It shows exterior and interior walls, doors, stairways, and mechanical equipment. The Floor Plan shows the floor as seen from above with the upper part of the building removed. *(See Figure 2.)*

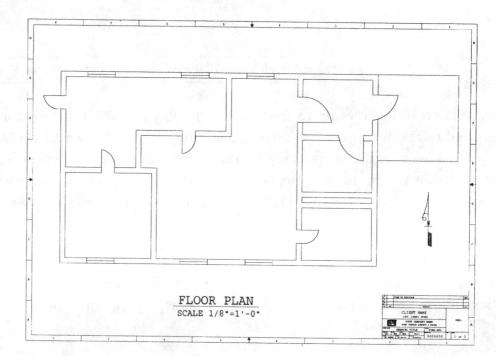

Figure 2. Floor Plan

FLOOR PLAN
SCALE 1/8"=1'-0"

The ROOF PLAN is a view of the roof from above a building. In *Figure 3* below, the arrows show the roof slopes towards the drains.

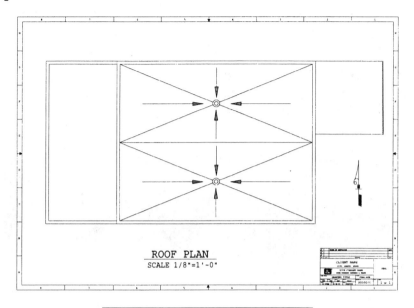

ROOF PLAN
SCALE 1/8"=1'-0"

Figure 3. Roof Plan (Top View)

1.2.3 Elevation Drawings

ELEVATION DRAWINGS *(11" x 17")* are side views of a building as seen by a person looking at each side. These views are called elevations because they show height. On a building drawing, there are standard names for different elevations. For example, the side of a building facing south is called the south elevation. Elevation drawings can show the exterior style of the building, as well as the placement of doors, windows, chimneys and decorative trim. *Figure 4* is a typical elevation drawing.

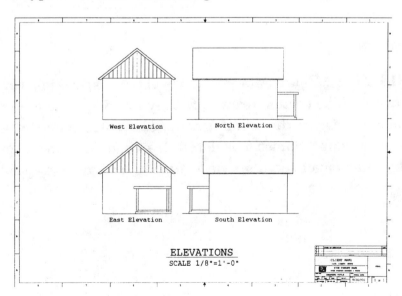

West Elevation   North Elevation

East Elevation   South Elevation

ELEVATIONS
SCALE 1/8"=1'-0"

Figure 4. Typical Elevation Drawing

1.2.4 Sectional Drawings

SECTIONAL DRAWINGS *(11" x 17")* are cutaway drawings that show the inside of an object or building. They are used to show the construction materials and how the parts of the object or building fit together. They are normally drawn to a larger scale than used in plan views. Next to the plan and elevation views, they are the most important drawings. The wall section in *Figure 5* is an example of a sectional view.

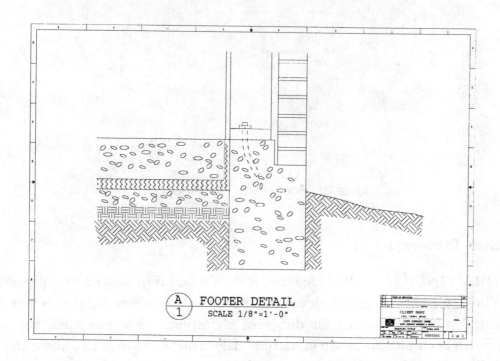

Figure 5. Sectional View (Wall Section)

1.2.5 Detail Drawings

DETAIL DRAWINGS *(11" x 17")* are enlarged views of some special features of a structure, such as windows or doors. Like section drawings, they are drawn to a larger scale in order to make the details clearer. Often the detail drawings are placed on the same sheet where the feature appears in the plan. Sometimes detail drawings are placed on separate sheets and are referenced by number on the plan view. *Figure 6* shows a typical detail drawing.

CORE CURRICULA TRAINEE TASK MODULE 00105

Basic Math

Module 00102

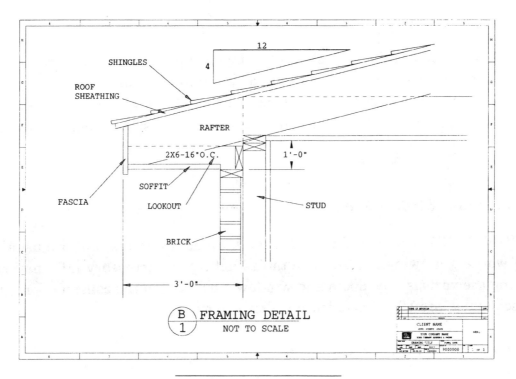

Figure 6. Detail Drawing

1.3.0 AUXILIARY DRAWINGS

Auxiliary (supplemental) Drawings consist of the following plans and drawings:

- Electrical Plans
- Plumbing Plan
- Foundation Plan
- HVAC Plan
- Door and Window Schedules
- Framing Plan

1.3.1 Electrical Plans

ELECTRICAL PLANS *(11" x 17")* may appear on the floor plan itself for construction jobs. For more complex structures, a separate plan, omitting unnecessary details and showing the electrical layout, is added to the set of plans. Included in this plan are: location of the meter, distribution panel, switches, convenience outlets, and special outlets (e.g., telephone). Conduit Plans (for conduit wiring) may be included on this plan.

1.3.2 HVAC Plan

The **HVAC** (Heating, Ventilation, Air Conditioning) PLAN *(11" x 17")*, like the electrical drawing, is included on the regular floor plan for simple air conditioning (heating and cooling) installations. For more complex jobs, a separate HVAC plan is added to the set of plans. Piping System Plans (for gas, oil, steam heat) may be included on this plan.

1.3.3 Plumbing Plan

The PLUMBING PLAN *(11" x 17")* shows the layout for the plumbing (water) system that supplies the hot and cold water, the sewage disposal system, and the location of plumbing fixtures. Plans for small residences may include the entire plumbing system plan on one drawing. For more complex jobs, separate plans for each system may be used.

1.3.4
Door And Window Schedules

DOOR AND WINDOW SCHEDULES *(11" x 17")* are not drawings, but are usually included in a set of working drawings. These are tables listing the necessary information and sizes for specifying the various type doors and windows included in the construction. On smaller jobs, the schedules are often included on plan sheets.

1.3.5 Foundation And Framing Plans

The FOUNDATION PLANS *(11" x 17")* include foundation and support walls, footings, **piers**, plumbing and electrical inlets/outlets (e.g., sewer drains, water pipes, electrical conduits, or bus cables). Like a Floor Plan, Foundation Plans show the lowest level of the building, including footings and foundations. Details and sectional drawings are sometimes included on this sheet. *Figure 7* provides an example of a Foundation Plan.

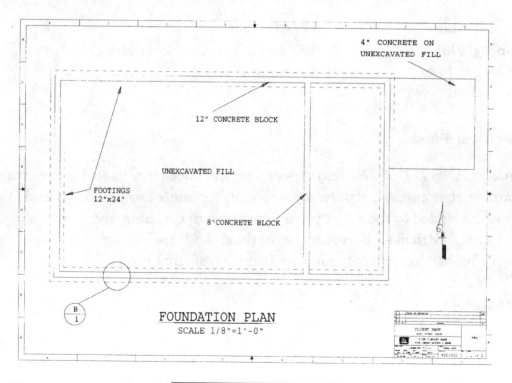

Figure 7. Foundation Plan

FRAMING PLANS are included for the framing of the roof, floors, and various elevations or wall sections. These plans are comprised of detailed cross-sections showing how the framing is to be constructed. The Foundation and Framing Plans may be included in other plans or may appear as a separate plan.

1.4.0   SELF-CHECK REVIEW / PRACTICE QUESTIONS 1

1. A blueprint is:

    A.   An architectural drawing of a blue building.
    B.   A computerized copy of large buildings.
    C.   A copy of a drawing in which, historically, the lines are white and the background is blue.
    D.   A tracing of a copy of a construction site.

2. Typically, the "players" on a construction job include:

    A.   Architects, engineers, bankers.
    B.   Architects, engineers, bankers, owners, workers.
    C.   Architects, engineers, bankers, owners.
    D.   Architects, astrologers, bankers, owners, workers.

3. Blueprints are also referred to as:

    A.   Working drawings.
    B.   Tracings.
    C.   Gridlines.
    D.   Outlines.

4. Why are blueprints important on a construction job?

    A.   They show how the building will look when completed.
    B.   They depict the contour lines of the site.
    C.   They illustrate how rooms compare to each other in size and building materials.
    D.   They form the basis of agreement and understanding among players.

Identify these plans and drawings. Write the plan/drawing name beside its number.

- Site Plan
- Elevation Drawing
- Detail Drawing
- Floor Plan
- HVAC Plan
- Door and Window Schedule

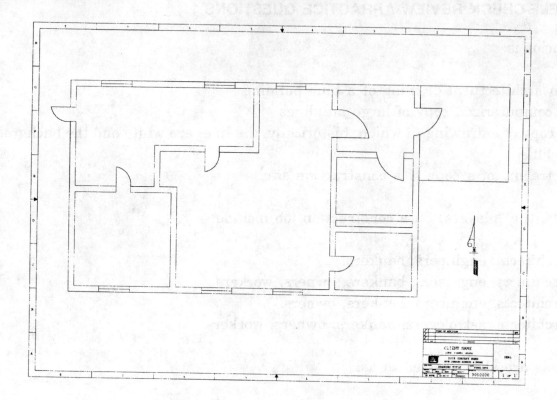

5: _____

Figure 8. Identify This Plan (Con't. on next page)

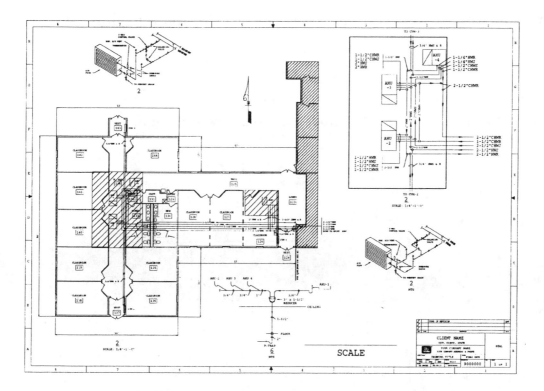

6: _____

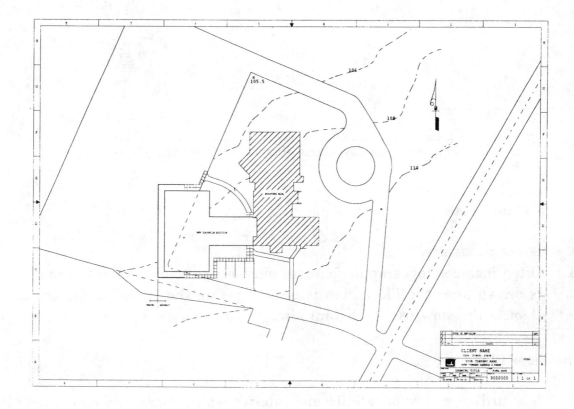

7: _____

Figure 8. Identify These Plans and Drawings (Con't. on next page)

(Continued) Identify these plans and drawings. Write the plan/drawing name beside its number.

- Site Plan
- Elevation Drawing
- Detail Drawing

- Floor Plan
- HVAC Plan
- Door and Window Schedule

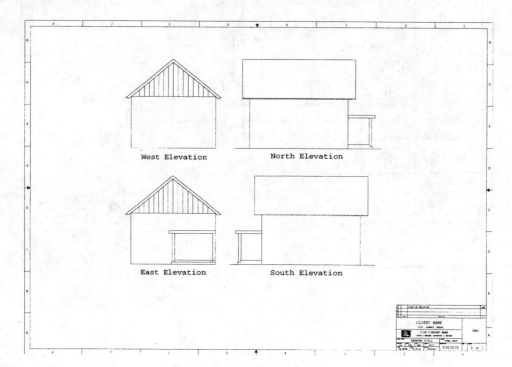

8. _____

Figure 9. Identify These Plans and Drawings (Con't.)

9. A Site Plan:

    A. Is not necessary.
    B. Often includes topographic features such as contour lines and trees.
    C. Is drawn after the Floor Plan is drawn, before the HVAC is designed.
    D. Depicts the view from a distant site.

10. The Floor Plan:

    A. Is a bird's-eye view of exterior and interior walls, doors, and stairways. Probably the most important drawing.
    B. Is a side view of interior walls, floors, and ceilings. Not a very important drawing.
    C. Shows how the floorboards fit together.
    D. Provides step-by-step instructions on how to lay a floor, including what tools to use.

11. Elevation Drawings:

    A.   Are views taken from a location above the building.
    B.   Show a 3-dimensional view of the building.
    C.   Show how the elevator will be installed in the building.
    D.   Are side views of the building or object.

12. Sectional Drawings:

    A.   Take a section of the building and place it over another section to form an overlay effect.
    B.   Show roof views.
    C.   Are cutaway drawings that show the inside of an object or building.
    D.   Show how the walls fit together.

13. Detail Drawings are:

    A.   Enlarged views of special features of a structure (e.g., windows or doors).
    B.   The details of the site, such as elevation, water tables, contour lines.
    C.   Illustrations and description of how the roof is constructed.
    D.   Show the details of the workers involved on the job, such as ages and skill level.

14. Door and Window Schedules:

    A.   Show the time-frame in which doors and windows will be installed.
    B.   Show who will be installing each of these items.
    C.   Are tables that list the cost of doors and windows.
    D.   Specify the types of doors and windows used in the construction.

15. On the Electrical Plan, you will find:

    A.   Location of fireplaces.
    B.   Location of meters, outlets.
    C.   Location of windows and doors.
    D.   Location of sewage lines.

16. HVAC is an abbreviation for:

    A.   Heating, Ventilation, and Air Conditioning.
    B.   Heating, Vaccination, and Air Conditioning.
    C.   Heating, Ventilation, and Airports.
    D.   Healing, Ventilation, and Air Conditioning.

17. The Plumbing Plan shows:

    A.   Location of fireplaces, stairways.
    B.   The layout of the plan that supplies wiring inlets/outlets, refrigeration plans, and air conditioning flows.
    C.   Location of electrical meters, outlets.
    D.   The layout for the plan that supplies water, the sewage disposal system, and the location of plumbing fixtures.

18. Framing Plans show:

    A.   The outline of the building.
    B.   The framing of the roof, floors, and various elevations or wall sections.
    C.   Where framed pictures will be placed, from the basement level to the top floor.
    D.   Where on the site the building will be placed.

19. Foundation Plans include:

    A.   The framing of the roof, floors, and various elevations or wall sections.
    B.   The outline of the building.
    C.   Foundation walls, footings, piers, fireplaces, stairways.
    D.   Where on the site the building will be placed.

20. The four basic views shown on construction drawings are:

    A.   Elevations, plan views, sections, and details.
    B.   Elevators, plan views, sections, and details.
    C.   Elevations, plan views, c-sections, and details.
    D.   Elevations, sections, and details.

2.0.0　COMPONENTS OF THE BLUEPRINT

All blueprints are laid out in a fairly standardized format. In this section, we will discuss five parts of the blueprint: (*See Figure 10 and Chart # 10*).

1. Title Block
2. Design Drawing Area
   - Drawing Title
   - North Arrow
   - Elevation
3. Legend
4. Revision Block
5. Scale

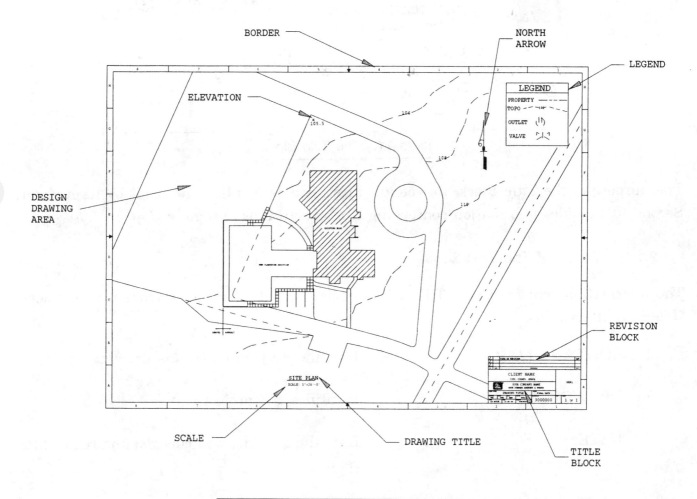

Figure 10. Components of the Blueprint

2.1.0   TITLE BLOCK

When you approach any blueprint, the first thing you must do is look at the Title Block. The Title Block is normally located in the lower right-hand corner of the blueprint or drawing.

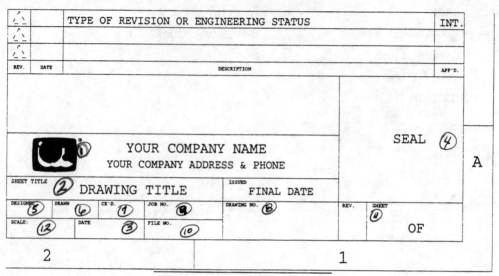

Figure 11.   The Title Block

The purpose of the Title Block is twofold. First, it aids in filing the print by its number. Second, it provides information concerning the structure or assembly.

2.1.1   Parts Of The Title Block

The information contained in a Title Block varies according to the architect or engineer. Generally it contains:

1. COMPANY LOGO   Usually pre-printed on the drawing.

2. DRAWING TITLE   Identifies the project.

3. DATE   Date the drawing was checked and readied for seal.

4. PROFESSIONAL STAMP/SEAL   Registered seal, should approval by Engineer or Architect be required.

5. DESIGN SUPERVISOR INITIALS   Person who designed the project.
   AND DATE

6.	DRAWN BY INITIALS	Person who drafted the drawings.
7.	CHECKED BY INITIALS	Person who reviewed the drawings.
8.	DRAWING NUMBER	Code numbers assigned to a project.
9.	CONTRACT/PROJECT NUMBER	Contract/project number assigned to a project.
10.	FILE NUMBER	Computer file number.
11.	SHEET NUMBER	Which sheet in the series of drawings.
12.	SCALE	The ratio between the size of an object being drawn and its actual size.
13.	REVISION BLOCK	Number, description, date, and initials of person making any revision. May be located as part of Title Block or near it.
14.	ENGINEERING STATUS	Depicts the status of the blueprint (e.g., "released for" information). Often appears in Revision Block.

Locate each of these in *Figure 11*.

If there is a diagonal or slanted line drawn across a block (not shown in *Figure 11*), it indicates that the information is not required or is provided elsewhere on the drawing.

Every company has its own numbering system for project numbers, departments, etc., as well as placement locations for the title and revision blocks. [Your instructor *will provide* and explain your company's system during class.]

2.2.0 DESIGN DRAWING AREA

The design drawing area is where the building plans are drawn. In this area, you will find the Drawing Title, North Arrow, and elevation, in addition to the Title Block and Legend. Symbols and lines are used to represent electrical, plumbing, and structural elements. *(See Figure 12.)*

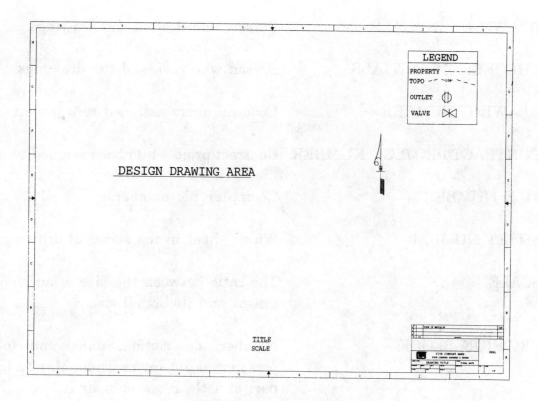

Figure 12. Design Drawing Area

Section 4.0.0 explains the various line types and symbols used on blueprints.

2.3.0 LEGEND

The legend is used to explain or define symbols or special marks placed on the drawing. *(See Figure 13.)*

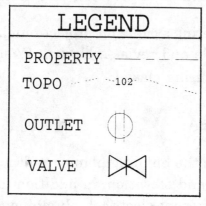

Figure 13. A Sample Legend

2.4.0 REVISION BLOCK

The Revision Block is located within the drawing area, usually in the lower right of the drawing within or near the Title Block. The location depends on the conventions of the designing company. It is used to record any changes (revisions) to the drawing, and typically contains the revision number, brief description, date, and initials of the person making any revisions. All revisions are noted in this block and are dated and identified by a letter or number. Usually, the most recent revision is circled.

WARNING!   It is essential to note the revision designation on a blueprint and to use only the latest issue; otherwise, costly mistakes will result.

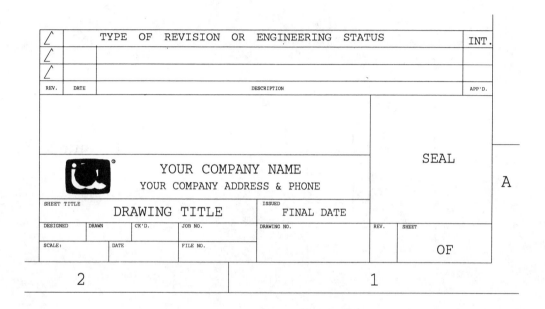

Figure 14. The Revision Block

2.5.0   SCALE

Construction projects are drawn to a reduced scale to make the blueprints useful on the job. The scale of a drawing is indicated in one of the spaces within the Title Block, beneath the drawing itself, or in both locations. It indicates the size of the drawing as compared to the actual size of the object represented.

The type of scale used on a drawing depends on the size of the objects being shown, the available space on the drawing paper, and the type of plan the scale is on.

Look at the scale on *Foldout 1, Site Plan*. It reads **SCALE - 1" = 20'- 0,"** which means that "every one inch on the drawing represents twenty feet, no inches." This is a typical scale for a Site Plan because Site Plans are developed by civil engineers. The scale used to develop Site Plans is an engineer's scale, described in Section 3.1.0. These scales are divided in 10, 20, 30, 40, 50, or 60 divisions of an inch.

Now look at the scale on *Foldout 2, Floor Plan*. It reads **SCALE - 1/4" = 1'- 0."** This is a typical scale used on plans other than the Site Plan, such as Floor Plans. It is developed by an architect, who uses an architect's scale, described in Section 3.2.0. These scales are divided into fractions of an inch. *(See Figure 15.)*

Other scales developed using an architect's scale are:

- Enlarged details: 1" = 1'-0"   1-1/2" = 1'-0"
- Sectional details: 1/2" = 1'-0"   3/4" = 1'-0" (some details may require full size)

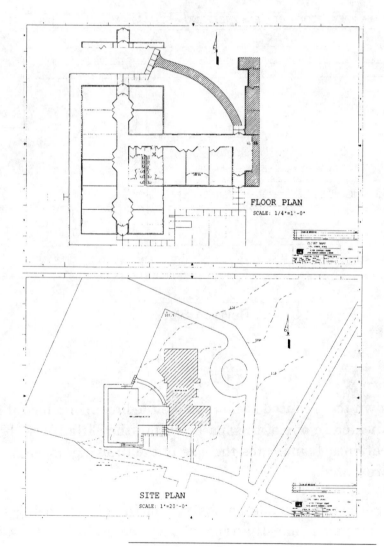

Floor Plan
(Architect's Scale)

Site Plan
(Engineer's Scale)

Figure 15.   Scales Used on Blueprints

Identify the parts of the blueprint. Select each option from the list on the left, locate its number on *Figure 16* below, and write the name in the corresponding blank.

4 • Title Block
1 • Legend
5 • Revision Block
8 • Design Drawing Area
3 • Scale
2 • Plan Title
7 • Elevation
6 • North Arrow

1. LEGEND
2. PLAN TITLE
3. SCALE
4. TITLE BLOCK
5. REVISION
6. NORTH ARROW
7. ELEVATION
8. DESIGN DRAWING AREA

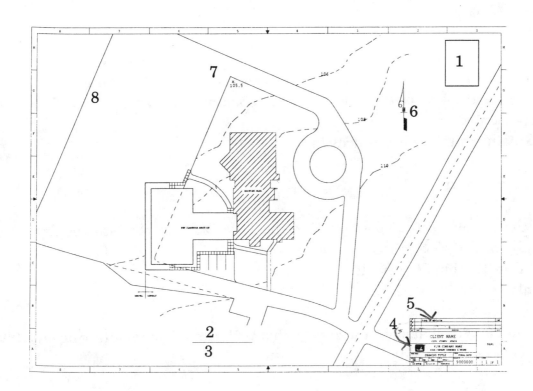

Figure 16. Identify the Parts of the Blueprint

9. The Title Block generally contains the following information:

A. Project Title and Number.
B. Drawing Title.
C. Sheet Number.
D. All of the above.

10. The Drawing Title is found in the Title Block and identifies:

    A.   Who made the drawing.
    B.   Where the project is located.
    C.   What is drawn on the sheet.
    (D.) The project.

11. The Legend is used to:

    A.   Tell the history of the building.
    (B.) Explain symbols on the drawing.
    C.   Define terms that are used only by electrical engineers.
    D.   Show the latest revision date.

12. The Scale is used to:

    A.   Weigh the hand tools and power tools used to construct the building.
    B.   Show a comparison between the various rooms in the building.
    (C.) Indicate the size of the drawing as compared to the actual size of the object represented.
    D.   Explain symbols on the drawing.

13. The Scale is shown:

    (A.) Either in one of the spaces within the Title Block or beneath the drawing itself.
    B.   Only on the Floor Plan.
    C.   Within the building itself.
    D.   Only in the Title Block.

14. If the actual length of an object is 4 feet, what is the length of the drawing on a blueprint with a scale of 1/4" = 1'-0" ?

    (A.) 2"
    B.   1"
    C.   1-1/2"
    D.   4"

15. Revisions to the drawing are entered in the Revision Block and contain:

    A.   Only the date of revision.
    B.   Project description, engineering approvals, intended date of completion.
    (C.) Revision number, description, date, initials of person making revision.
    D.   Customer approval.

16. How can you tell what the latest revision of a blueprint is?

A. It's in red.
B. It's circled.
C. It has an "X" through it.
D. It's written larger than the other revisions.

3.0.0   MEASURING TOOLS

Scales are also the term used for the measuring tools used to draw or measure the lines of a blueprint drawing. The three types of scales used to measure are:

- The engineer's scale
- The architect's scale
- The metric scale

3.1.0   THE ENGINEER'S SCALE

The engineer's scale is divided into decimal graduations (10, 20, 30, 40, 50 and 60 divisions to the inch). These scales are used for plotting and map drawing and in the graphic solution of problems such as survey and site plans. *(See Figure 17.)*

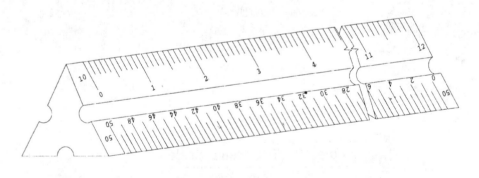

Figure 17. The Engineer's Scale

3.2.0   THE ARCHITECT'S SCALE

The architect's scale is used on all plans other than Site Plans. They are divided into proportional feet and inches. The triangular form is used a great deal in schools and in some drafting offices because it contains a variety of scales on a single scale (1/8," 1/4," 1/16," 1/32," etc.). It can be read either from left to right or from right to left. *(See Figure 18.)*

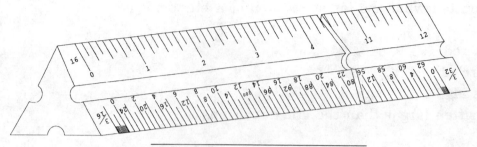

Figure 18. The Architect's Scale

3.3.0   THE METRIC SCALE

Metric scales *(Figure 19)*, are divided into centimeters (cm), with the centimeters divided into 10-divisioned millimeters (mm), or into 20-divisioned half millimeters. Some scales are made with metric divisions on one edge and inch divisions on the opposite edge. Many companies that deal in international trade have adopted a dual-dimensioning system expressed in both metric and English symbols. If you recall from the Math Module, metric will be the required scale for companies who want to work on government contracts.

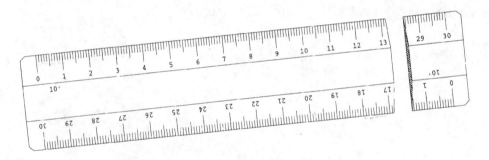

Figure 19. The Metric Scale

3.4.0   SELF-CHECK REVIEW / PRACTICE QUESTIONS 3

1.   The three types of scales used to draw or measure the lines of a blueprint drawing are:

    A.   Architect's scale, engineer's scale, metered scale
    B.   Architect's scale, engineer's scale, metric scale
    C.   Archie's scale, engineer's scale, metric scale
    D.   Architect's scale, electrician's scale, metric scale

Identify each of these scales (write the letter of the pictured scale by its name in *Figure 20*.)

2. Architect's scale: ___C___
3. Engineer's scale: ___A___
4. Metric scale: ___B___

Scale A

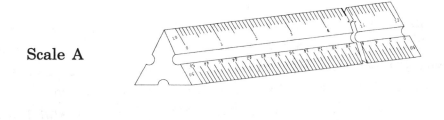

Scale B

Scale C

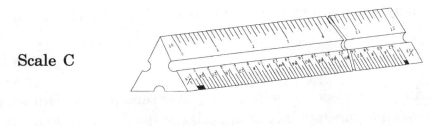

Figure 20. Identify the Measuring Scales

Match the definition to its proper term. Fill in the letter beside its name.

5. Metric scale: ___a___   a. Used in the graphic solution of problems such as survey and site plans

6. Architect's scale: ___c___   b. Divided into centimeters and millimeters

7. Engineer's scale: ___b___   c. Divided into proportional feet and inches

The language of blueprint reading is written in lines and symbols that appear on drawings. Drafters use a definite system of lines and symbols. The meaning of each line and symbol is presented in this section.

You will find that many construction companies require that you have at least a familiarity with the symbols used in blueprints before they will hire you. Each trade has its own symbols, and the craftworker of each trade should learn to recognize the symbols for all other trades. For example, the electrician should understand the carpenter's symbols, the carpenter should understand the plumber's symbols, etc. In this way, each craftworker will know what obstacles may be encountered while working and will then be better prepared to cope with them.

A designer may sometimes wish to use symbols other than those recommended for a particular specialty, in which case the symbols should be in a legend showing what they mean.

The following section will help you learn to read and interpret blueprint lines and symbols.

4.1.0   LINE TYPES

Have you ever heard of radio Morse Code? If sailors wanted to send a "Save Our Ship" (SOS) emergency message, they would transmit this as a series of dots and dashes.

Dot-dot-dot, dash-dash-dash, dot-dot-dot

Builders have their own Morse Code for their trade. These are the line types commonly used on construction drawings and are referred to as the "Alphabet of Lines." A description of each follows *Figure 21.*

Note   These lines will vary slightly throughout various companies.

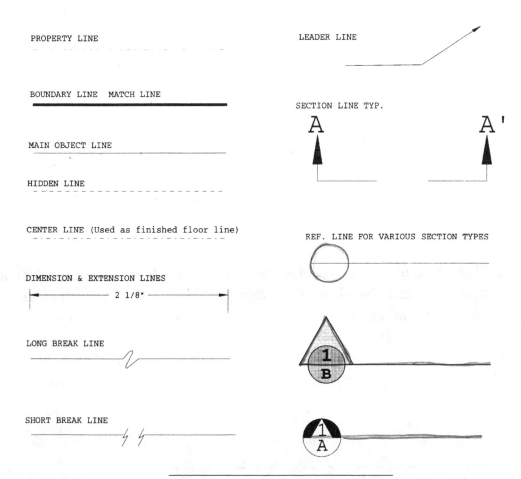

Figure 21. The Alphabet of Lines

1. PROPERTY LINE. The property line is an extra heavy line with long dashes alternating with two short dashes. It indicates land boundaries on the site plan. *(See Figure 22.)*

Figure 22. Property Line

2. BOUNDARY LINE (Match Line). Boundary lines are used to show where drawings that are continued on other sheets match up. It's as if you cut the prints with scissors and match the "match" lines together to complete the puzzle. *(See Figure 23.)*

Figure 23. Boundary Line (Match Line)

3. MAIN OBJECT LINE. Object lines represent the main outline of the features of the object, building, or wall. The object line is a heavy, continuous line, showing all edges and surfaces. (*See Figure 24.*)

_____

Figure 24. Main Object Line

4. HIDDEN LINES. Hidden lines are used to represent edges of an object that would normally be hidden behind the view of an object or are continued on another drawing. Medium weight, short dashes represent these invisible lines. The worker must look for another view in the set of drawings to locate at what location these edges occur. Often, these hidden parts will be revealed in an elevation or in a sectional view. Hidden lines are used only when their presence helps to clarify a drawing. (*See Figure 25.*)

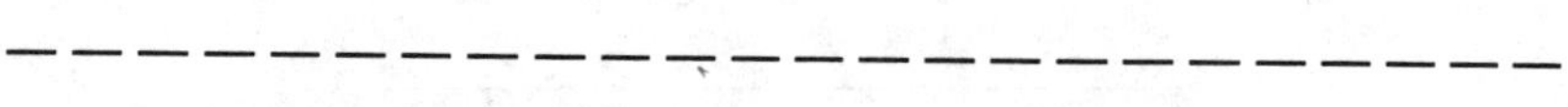

— — — — — — — — — — — — — — — — —

Figure 25. Hidden Lines

5. CENTER LINE. The center line is used to indicate centers of objects such as columns, equipment, and fixtures. This line is light in weight and composed of alternating long and short dashes. It is often labeled with a ℄ . (*See Figure 26.*)

— — — — — — — — — — — — — —

Figure 26. A Center Line

6. DIMENSION AND EXTENSION LINES. The size of an object is indicated by DIMENSION LINES on the drawing. Blueprints are usually dimensioned in inches. Size **dimensions** are placed above the dimension lines. Some companies place the dimension between separate arrowhead-tipped lines. Normal practice is to put dimensions outside the object's outline; however, at times it may be necessary to put dimensions inside the outline. In either case, dimensions should be put in clear view on the drawing. (*See Figure 27.*)

EXTENSION LINES are lightweight and extend from the edge of the outline. They start about 1/16-inch away from the view outline. Extension lines help to explain dimension indications on drawings.

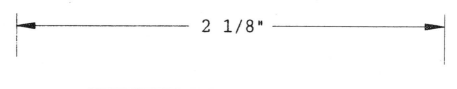

Figure 27.   Dimension and Extension Lines

7. BREAK LINES. Break lines are used to indicate that an object has been broken off at that point to save space. When the break on the drawing is lengthy, LONG BREAK lines with freehand "zig-zags" are used. SHORT BREAK lines are used when the break is short, such as across a joist or beam. This is a thick line drawn freehand. Some architectural draftsmen use the long break line for all drafts. (See Figure 28.)

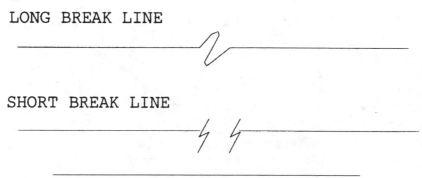

Figure 28.   Long Break and Short Break Lines

8. SECTION LINE. Section lines, sometimes referred to as cross-hatch lines, are thin lines usually drawn at an angle of 45 degrees. (See Figure 29) They are used to indicate a cross-section of material. They are marked with arrows to show how the cut is being made. A letter or other designation is usually drawn above each arrow. The section is then referred to as "Section A-A," for example.

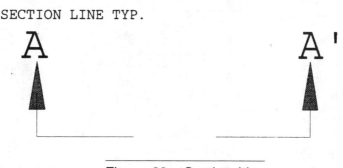

Figure 29.   Section Line

9. **REFERENCE LINE FOR SECTION.** Reference lines are solid lines indicating that an imaginary cut has been made at this point and that a detail section is shown elsewhere on the drawings. The arrow indicates the direction in which the section is viewed. The letters and numerals, usually in a circle attached to the reference line, indicate the particular section and where it will be found. *(See Figure 30.)*

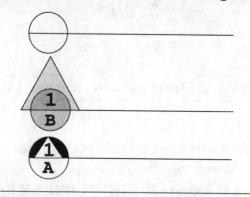

Figure 30. Reference Line for Section

10. **LEADER LINES.** These lines point to specific objects or areas on the drawing. They indicate where explanatory notes or dimensions are to be used. *(See Figure 31.)*

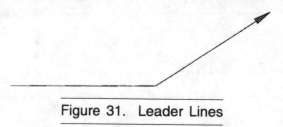

Figure 31. Leader Lines

Figure 32 illustrates examples of some of these lines. See if you can identify the line types.

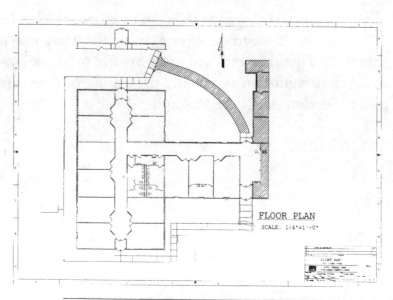

Figure 32. Blueprint Showing Line Types

4.2.0 SYMBOLS

In addition to the various lines that give meaning to a drawing, a number of symbols (sometimes referred to as conventions) are commonly used on construction drawings. Like the alphabet of lines, each company may have variations on their use of specific symbols or use symbols unique to that company. Always refer to the legend for a description of symbols used on a drawing.

Four types of symbols are used:

- Building Material Symbols such as steel, wood, concrete, or gravel (*Figure 33*)
- Electrical Symbols (*Figure 34*)
- Piping Symbols (*Figure 35*)
- Door and Window Symbols (*Figure 36*)

All craftworkers should be familiar with these basic symbols. The symbols specific to each craft area will be explained in more detail in the advanced craft study modules. For quick reference, these symbols are also found in Appendix C.

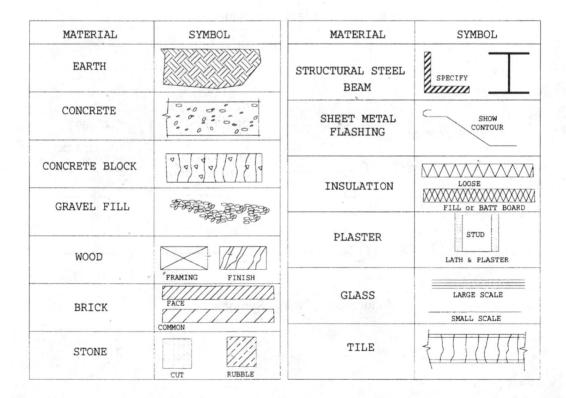

Figure 33. Building Material Symbols

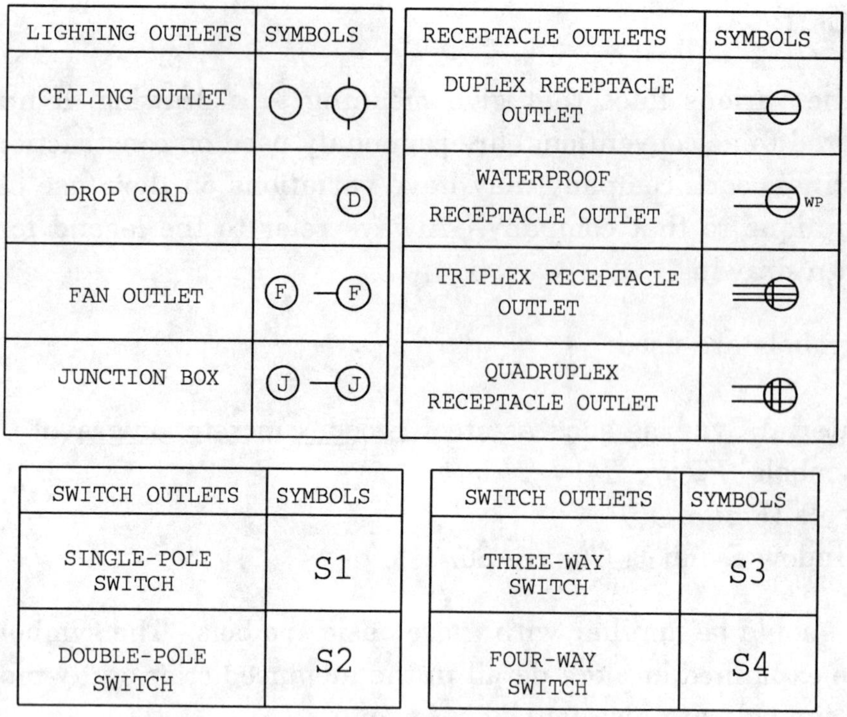

LIGHTING OUTLETS	SYMBOLS	RECEPTACLE OUTLETS	SYMBOLS
CEILING OUTLET	○ ⊕	DUPLEX RECEPTACLE OUTLET	⊖
DROP CORD	Ⓓ	WATERPROOF RECEPTACLE OUTLET	⊖ WP
FAN OUTLET	Ⓕ ─Ⓕ	TRIPLEX RECEPTACLE OUTLET	⊜
JUNCTION BOX	Ⓙ ─Ⓙ	QUADRUPLEX RECEPTACLE OUTLET	⊞

SWITCH OUTLETS	SYMBOLS	SWITCH OUTLETS	SYMBOLS
SINGLE-POLE SWITCH	S1	THREE-WAY SWITCH	S3
DOUBLE-POLE SWITCH	S2	FOUR-WAY SWITCH	S4

Figure 34. Electrical Symbols

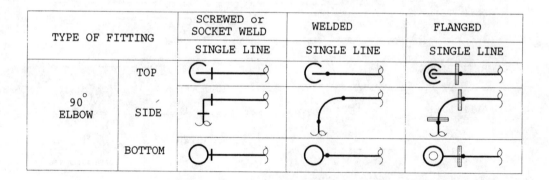

TYPE OF FITTING		SCREWED or SOCKET WELD	WELDED	FLANGED
		SINGLE LINE	SINGLE LINE	SINGLE LINE
90° ELBOW	TOP			
	SIDE			
	BOTTOM			

Figure 35. Piping Symbols

DOOR TYPE	SYMBOL	WINDOW TYPE	SYMBOL
SINGLE SWING		AWNING	
SLIDER		FIXED SASH	
BIFOLD		DOUBLE HUNG	
FRENCH		CASEMENT	
ACCORDION		HORIZONTAL SLIDER	

Figure 36. Door and Window Symbols

Some of the symbols you learned about are pictured in *Figure 37*. See if you can identify them.

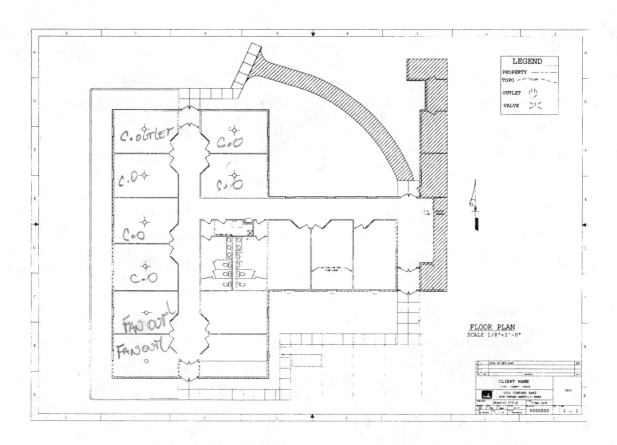

Figure 37. Can You Identify the Symbols on this Floor Plan?

The following *(Figure 38)* are standard abbreviations used in blueprint drawings.

ADD.	addition		N	north
AGGR	aggregate		NO.	number
L ⟶	angle		OC	on center
B	bathroom		OPP	opposite
BR	bedroom		O.D.	outside diameter
BM	bench mark		PNL	panel
BRKT	bracket		PSI	pounds per square inch
CLK	caulk		PWR	power
CHFR	chamfer		REINF	reinforce
CND	conduit		RH	right-hand
CU FT	cubic foot, feet		RFA	released for approval
DIM.	dimension		RFC	released for construction
DR	drain		RFD	released for design
DWG	drawing		RFI	released for information
ELEV	elevation		SHTHG	sheathing
ESC	escutcheon		SQ	square
FAB	fabricate		STR	structural
FLGE	flange		SYM	symbol
FLR	floor		THERMO	thermostat
GR	grade		TYP	typical
GYP	gypsum		UNFIN	unfinished
HDW	hardware		VEL	velocity
HTR	heater		WV	wall vent
" or IN.	inch, inches		WHSE	warehouse
I.D.	inside diameter		WH	weep hole
LH	left-hand		WDW	window
MEZZ	mezzanine		WP	working pressure
MO	masonry opening			
MECH	mechanical			

Figure 38. Standard Abbreviations Used in Blueprint Drawings

1. It is important to be able to recognize the basic symbols for not only your trade area, but for others as well.

    (A) True.
    B. False.

2. Reference Lines for a Section:

    A. Are used to indicate a cross-section of material.
    B. Are used to indicate that an object has been removed from the drawing to save space.
    C. Are used to indicate centers of objects such as columns, equipment and fixtures.
    (D) Are solid lines indicating that an imaginary cut has been made, and that a detailed section is shown elsewhere on the drawing.

3. What is the "Alphabet of Lines?"

    (A.) The line-types commonly used on construction drawings.
    B. Lines that are indicated using letters of the alphabet.
    C. Lines that are used to show where pages match up when a large page has been broken   down to several smaller pages.
    D. Lines that indicate land boundaries on the site plan.

4. Main object lines represent the main outline of the features of the object, wall, or building.

    (A) True.
    B. False.

Match the name of the line type *(in Figure 39)* from the list below. Write the line type beside its number.

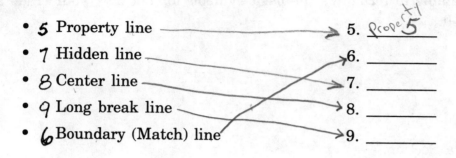

- **5** Property line   5. _Property **5**_
- **7** Hidden line   6. _____
- **8** Center line   7. _____
- **9** Long break line   8. _____
- **6** Boundary (Match) line   9. _____

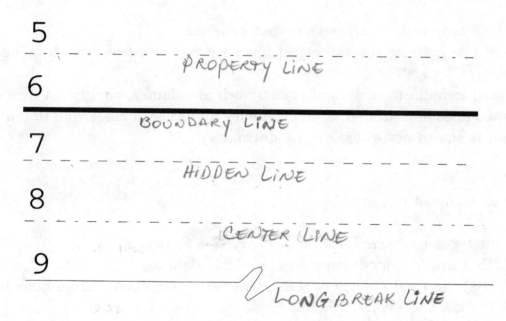

5   PROPERTY LINE

6   BOUNDARY LINE

7   HIDDEN LINE

8   CENTER LINE

9   LONG BREAK LINE

Figure 39. What are These Line Types?

Can you identify the symbols in *Figure 40*? Choose from the following list. Write the name of the symbol beside its number.

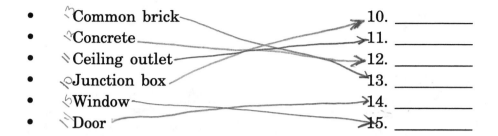

- Common brick
- Concrete
- Ceiling outlet
- Junction box
- Window
- Door

10. _____
11. _____
12. _____
13. _____
14. _____
15. _____

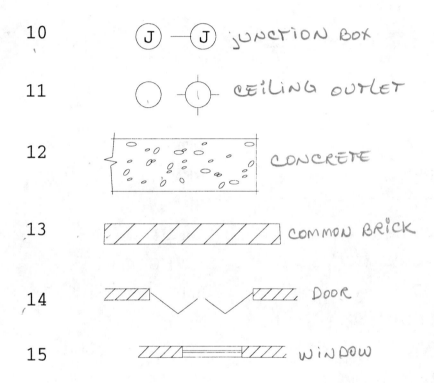

10   J — J JUNCTION BOX

11   ○ ○ CEILING OUTLET

12   CONCRETE

13   COMMON BRICK

14   DOOR

15   WINDOW

Figure 40. What Are These Symbols?

What are the abbreviations for the words below? Choose from those listed in *Figure 41*. Write the abbreviation beside its name.

16. Bathroom   B

17. Elevation   ELEV

18. Floor   FLR

19. Inch   " or IN.

20. Number   NO.

21. Power   PWR

22. Unfinished   UNFIN

23. Window   WDW

100% ☺

ADD. ADDITION	HTR HEATER	WP working Pressure
FLR FLOOR	RFA Realease for approval	SYM Symbol
O.D. OUTSIDE DIAM	BR BEDROOM	ELEV ELEVATION
WHSE WAREHOUSE	" or IN. INCH	THERM THERMO
AGGR Agregate	SHTHG SHEATING	ESC Escutcheon
GR Grade	BM Bench mark	NO. NUMERO
PNL PANEL	N NORTH	TYP Typical
WH weephole	I.D. INSIDE DIAM.	FAB Fabricate
L ANGEL	SQ SQUARE	OC On Center
HDW Hardware	CLK CAULK	UNFIN UNFINISHED
PWR POWER	LH. LEFT HAND	FLGE FLANGE
WDW WINDOW	ST	OPP opposite
B BATHROON	CU FT cubic Foot (Feet)	WV WALL VENT

Figure 41. Select Abbreviations from This List

1 3/3
39
-1 2 → wrong AN.
27 → correct AN.

CORE CURRICULA TRAINEE TASK MODULE 00105

6.0.0 USING GRID LINES TO IDENTIFY PLAN LOCATIONS

Have you ever used a street map to locate a street in a neighborhood? The map most likely used a grid to make locating a detailed area easier. The index referred you to "Page 24, E-7." You turned to page 24 of the map, located "E" along the side of the page, "7" along the top of the page, located the intersection of the two, and found your street.

Like the map example, the grid line system *(Figure 42)* is normally used for such identification. On a drawing, such as a floor plan, an imaginary grid divides the area into small parts called **bays**. The grid pointers are delineated in the border area.

By convention, the numbering and lettering system begins in the upper left-hand corner of the floor plan. The numbers are normally assigned horizontally; the letters, vertically. Several letters and a number have been eliminated from the system to avoid confusion.

- The Letters Omitted are: I, O, Q
- Number Omitted: 0

The use of column lines permits anyone familiar with the system to locate any point exactly.

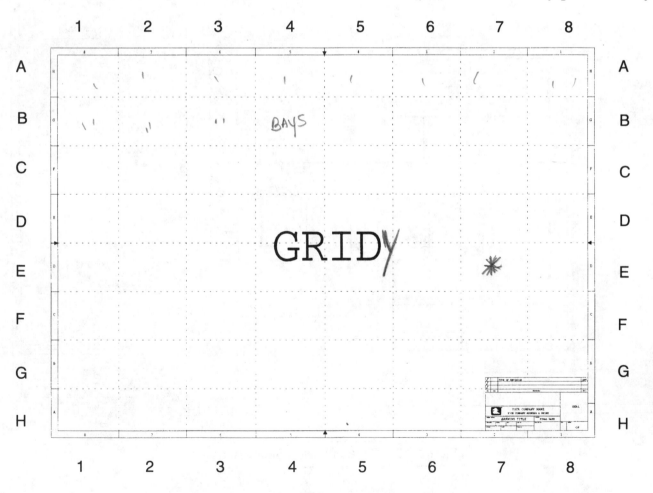

Figure 42. Column Line Grid

Dimensions are included on a drawing to show the size (in inches and fractions; feet and inches; inches and decimals; or millimeters, if metric is required) and location (by placement or arrow) of all parts within the blueprints. To do accurate work, you need to know how to read the dimensions from the drawings.

WARNING!   Always use the drawing's written dimensions; never measure a drawing or blueprint. Written dimensions are more accurate than those measured from the drawings, as the drawing could have been shrunk or stretched without your knowledge.

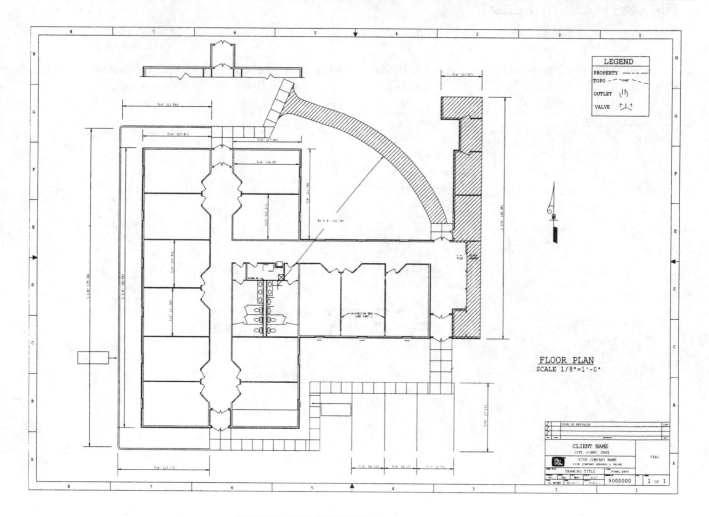

Figure 43. Blueprint with Dimensions

The most common means of terminating dimension lines are the arrowheads pictured in *Figure 43* above.

7.1.0   DIMENSIONING PRACTICES ON FLOOR PLANS

Exterior walls of solid masonry construction are dimensioned to the exterior surface (*see Figure 44*). Dimensions for exterior walls of frame and brick-veneer buildings usually start at the outside surface of the stud wall. Some architects show the dimension to the outside of the masonry on brick-veneer as well.

Interior walls are usually dimensioned to the center of partitions (*Figure 45*). However, some architects follow the practice of dimensioning to partition surfaces, then dimensioning the thickness of each partition.

Window and door openings are located by dimensions to their center lines (*Figure 46*). Doors or windows in narrow areas may not be dimensioned for location since it is obvious they would be centered in the space available. A reference symbol may be used to refer you to a door or window schedule located on a separate drawing.

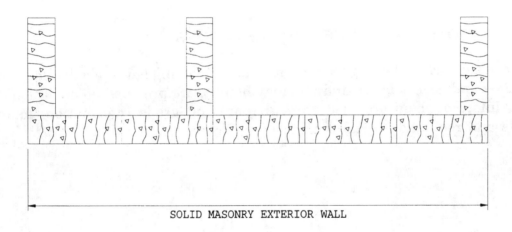

SOLID MASONRY EXTERIOR WALL

Figure 44.   Dimensioning Practices on Floor Plan Exterior Walls

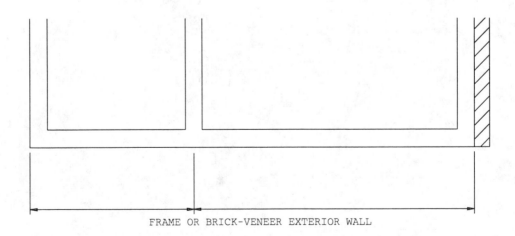

FRAME OR BRICK-VENEER EXTERIOR WALL

Figure 45.   Dimensioning Practices on Floor Plans Interior Walls

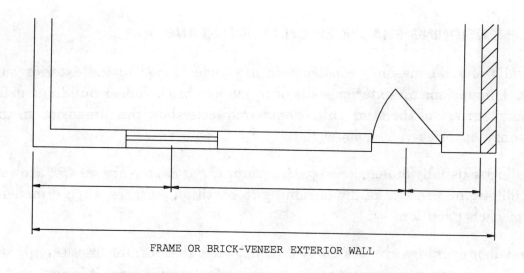

FRAME OR BRICK-VENEER EXTERIOR WALL

Figure 46. Dimensioning Practices on Floor Plans, Windows, and Doors

7.2.0   DIMENSIONING PRACTICES ON ELEVATIONS

Height dimensions, such as footing thickness, depth of footing below grade, floor and ceiling heights, window and door heights, and chimney height, are provided on elevation drawings. Additionally, information is provided through notes on grade information, materials for exterior walls and roof, and special details. *(See Figure 47.)*

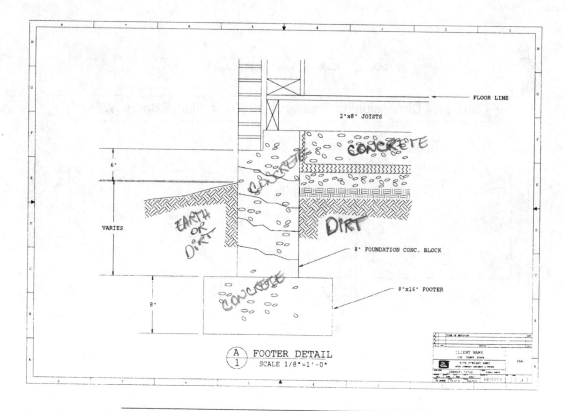

Figure 47. Dimensioning Practices on Elevations

CORE CURRICULA TRAINEE TASK MODULE 00105

7.3.0 DIMENSIONING PRACTICES ON SECTIONS AND DETAILS

For greater clarity, section and detail drawings frequently are drawn to a larger scale than plan and elevation drawings. Detail dimensions showing thicknesses of finished and sub-floor materials, joist sizes, molding location, etc. provide essential construction information. *(See Figure 48.)*

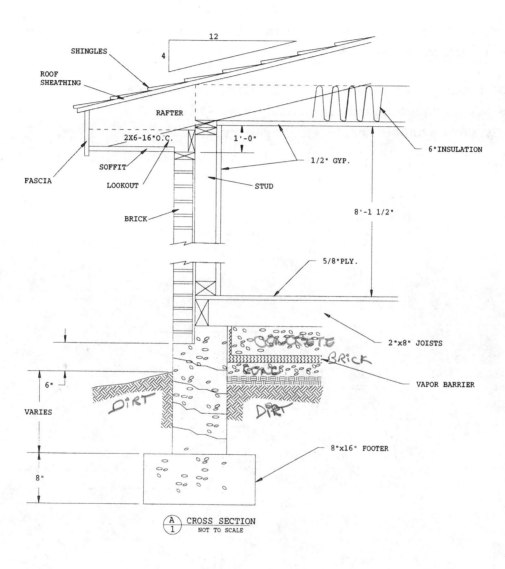

Figure 48. Dimensioning Practices on Sections and Details

8.0.0　WHAT IS COMPUTER-AIDED DESIGN (CAD)?

The use of computers is a cost effective method of increasing design and drafting productivity in mechanical, electronic, electrical, architectural, and civil engineering applications. Computer-aided design can increase productivity up to ten times with some of the systems that are available today.

A CAD system *(Figure 49)* generates drawings from computer programs. Advantages of the CAD over hand-drawn construction drawings are:

* it can be learned and used easily by drafters and designers,
* changes can be made quickly and easily, and
* commonly used symbols can be easily retrieved.

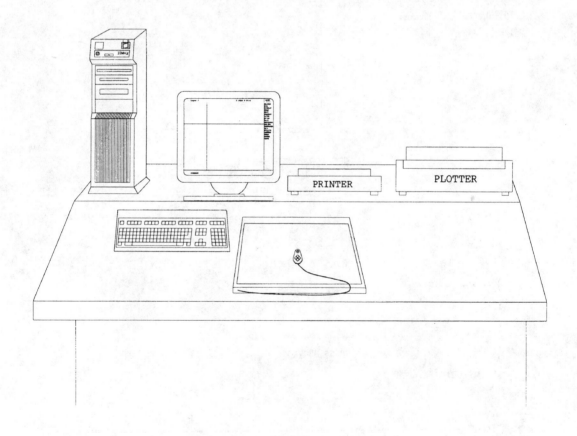

Figure 49. A Computer-Aided Design (CAD) System

9.0.0 CARE OF BLUEPRINTS

Blueprints are valuable records and must be cared for. Observe the following rules when handling blueprints:

1. Never write on a blueprint without authorization.
2. Keep blueprints clean. Dirty blueprints are hard to read and contribute to errors.
3. Fold and unfold blueprints carefully to avoid tearing.
4. Do not lay sharp tools or pointed objects on blueprints.
5. Fold blueprints so that the Title Block is visible.
6. Keep blueprints away from moisture.

10.0.0 STANDARD VS. *YOUR* COMPANY'S PROCEDURES

Although a standardized approach has been developed for reading and using blueprints, your company may have determined some specific ways to denote elements of a blueprint. At this time, your instructor will address those conventions that are unique to your company.

11.0.0 SELF-CHECK REVIEW / PRACTICE QUESTIONS

1. What are grid lines used for?

   A. To indicate a cross-section of material.
   B. To indicate that an object has been broken off at that point to save space.
   C. To indicate land boundaries on the site plan.
   D. To call out specific locations within a building or structure.

2. What do dimensions show?

   A. The scale of the parts.
   B. The size and location of all parts within the blueprint.
   C. How the walls fit together.
   D. The list of builders and contractors on a job.

3. Which is more accurate?

   A. Written dimensions.
   B. Dimensions measured from a drawing.

4. CAD is an abbreviation for:

    A.   Compruter-Activated Design.
    B.   Computer-Aided Design.
    C.   Computer-Aided Details.
    D.   Console-Aided Design.

5. It is O.K. to write on a blueprint without authorization.

    A.   True.
    B.   False.

6. Fold blueprints so that the _____ is visible.

    A.   Scale.
    B.   Legend.
    C.   Contractor's name.
    D.   Title Block.

7. All companies use the same standardized approach to reading and using blueprints.

    A.   True.
    B.   False.

SUMMARY

This module presented you with an *Introduction to Blueprints*. You should now have an overview of the following:

- **INTRODUCTION TO BLUEPRINTS**

  <u>Working Drawings</u>

  1. The Site Plan
  2. Plan Views (Floor Plan, Roof Plan)
  3. Elevation Drawings
  4. Sectional Drawings
  5. Detail Drawings

  <u>Auxiliary Drawings</u>

  1. Electrical Plan
  2. HVAC Plan
  3. Plumbing Plan
  4. Door and Window Schedules
  5. Foundation and Framing Plans

- **COMPONENTS OF THE BLUEPRINT**

  Title Block   Revision Block
  Design Drawing Area   Legend
  Scale

- **MEASURING TOOLS**

- **LINE TYPES AND SYMBOLS**

- **ABBREVIATIONS**

- **USING GRID LINES TO IDENTIFY PLAN LOCATIONS**

- **DIMENSIONS**

- **WHAT IS COMPUTER-AIDED DESIGN (CAD)?**

- **CARE OF BLUEPRINTS**

- **STANDARD VS. *YOUR* COMPANY'S PROCEDURES**

REFERENCES

For advanced study of topics covered in this Task Module, the following work is suggested:

Blueprint Reading for Construction, Goodheart-Willcox Co., South Holland, IL, 1987.

PERFORMANCE / LABORATORY EXERCISES

Identify the plans and drawings. Write the plan/drawing name beside its correct number in *Figure 50,* choosing from the list below.

- Site Plan
- Elevation Drawing
- Detail Drawing

- Floor Plan
- HVAC Plan
- Window and Floor Schedules

1: *ELEVATION DRAWING*

3: *FLOOR PLAN*

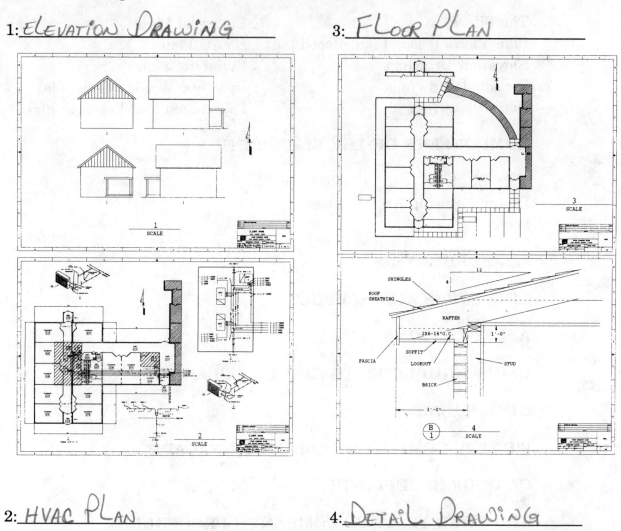

2: *HVAC PLAN*

4: *DETAIL DRAWING*

Figure 50. Identify These Plans and Drawings

Identify the parts of the blueprint *(Figure 51)*. Select your options from the list on the left to fill in the corresponding blanks to the right.

- Title Block
- Legend
- Revision Block
- Design Drawing Area
- Scale
- Plan Title

5. _LEGEND_
6. _PLAN TITLE_
7. _DESIGN DRAWING AREA_
8. _TITLE BLOCK._
9. _REVISION BLOCK_
10. _SCALE_

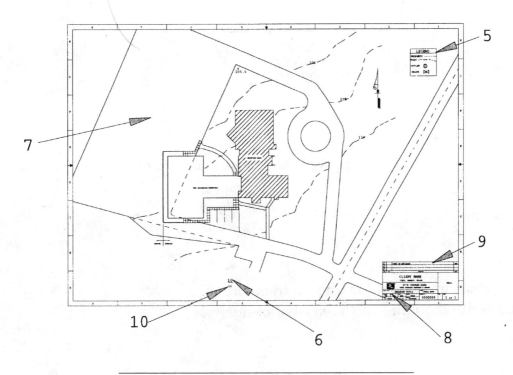

Figure 51. Identify the Parts of the Blueprint

Match the name of the line type from the list below. Write the line type *(Figure 52)* beside its letter.

- Property line
- Hidden line
- Center line
- Break line
- Boundary (Match) line

11. PROPERTY LINE
12. BOUNDARY LINE
13. HIDDEN LINE
14. CENTER LINE
15. BREAK LINE

Figure 52. What Are These Line Types?

Identify the symbols in *Figure 53*. Choose from the following list. Write the name of the symbol beside its number.

Common brick
Concrete
Ceiling outlet
Junction box
Window
Door

16. CONCRETE
17. COMMON BRICK
18. DOOR
19. WINDOW
20. CEILIN OUTLET
21. JUCTION BOX

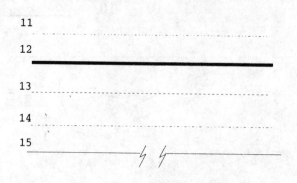

Figure 53. What Are These Symbols?

What are the abbreviations for the words below? Choose from those listed in *Figure 54*. Write the abbreviation beside its name.

22.   Addition   _ADD_
23.   North   _N_
24.   On Center   _OC_
25.   Panel   _PNL_
26.   Square   _SQ_
27.   Symbol   _SYM_
28.   Wall Vent   _WV_
29.   Working Pressure _WP_

ADD.	ELEV	" or IN.	O.D.	THERMNO
AGGR	ESC	I.D.	PNL	TYP
L	FAB	LH	PWR	UNFIN
B	FLGE	WP	RFA	WV
BR	FLR	N	SHTHG	WHSE
BM	GR	NO.	SQ	WH
CLK	HDW	OC	STR	WDW
CU FT	HTR	OPP	SYM	WP

Figure 54. Select Abbreviations from this List

30.   Ask the instructor for a full-size blueprint. Fold it correctly and pass it back to the instructor.

APPENDIX A: ANSWERS TO SELF-CHECK REVIEW / PRACTICE QUESTIONS

QUESTIONS 1.4.0

1. C
2. B
3. A
4. D
5. Floor Plan
6. HVAC Plan
7. Site Plan
8. Elevation Drawing
9. B
10. A
11. D
12. C
13. A
14. D
15. B
16. A
17. D
18. B
19. C
20. A

QUESTIONS 2.6.0

1. Legend
2. Plan Title
3. Scale
4. Title Block
5. Revision Block
6. North Arrow
7. Elevation
8. Design Drawing Area
9. D
10. D
11. B
12. C
13. A
14. B
15. C
16. B

QUESTIONS 3.4.0

1. B
2. C
3. A
4. B
5. B
6. C
7. A

QUESTIONS 5.1.0

1. A
2. D
3. A
4. A
5. property line
6. boundary (match) line
7. hidden line
8. center line
9. long break line
10. junction box
11. ceiling outlet
12. concrete
13. common brick
14. door
15. window
16. B
17. ELEV
18. FLR
19. " or IN.
20. NO.
21. PWR
22. UNFIN
23. WDW

QUESTIONS 11.0.0

1. D
2. B
3. A
4. B
5. B
6. D
7. B

ALPHABET OF LINES

PROPERTY LINE

BOUNDARY LINE   MATCH LINE

MAIN OBJECT LINE

HIDDEN LINE

CENTER LINE (Used as finished floor line)

DIMENSION & EXTENSION LINES

|←————— 2 1/8" —————→|

LONG BREAK LINE

SHORT BREAK LINE

LEADER LINE

SECTION LINE TYP.

A   A'

REF. LINE FOR VARIOUS SECTION TYPES

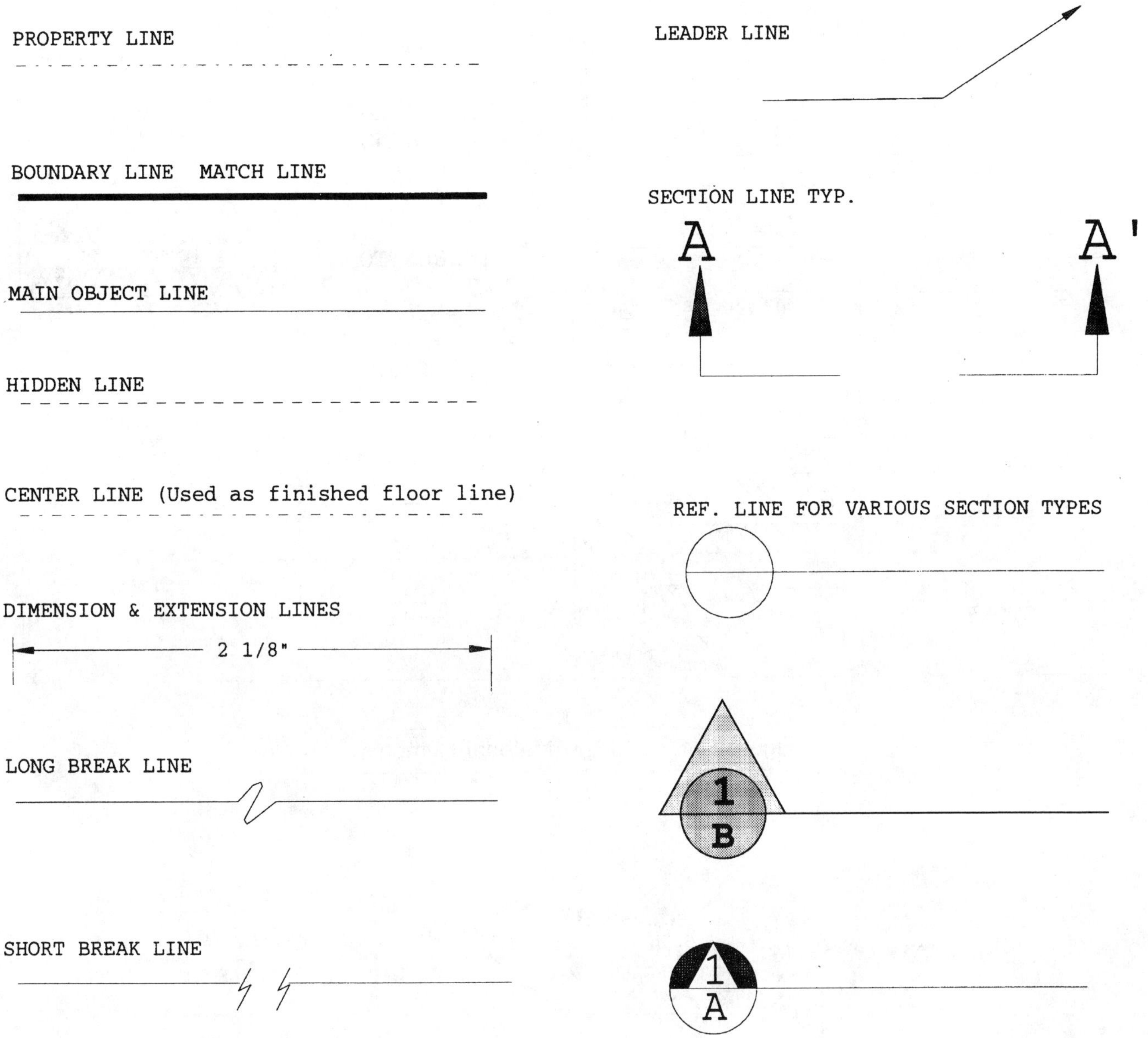

Figure B-1. Alphabet of Lines

APPENDIX C: CONSTRUCTION SYMBOLS

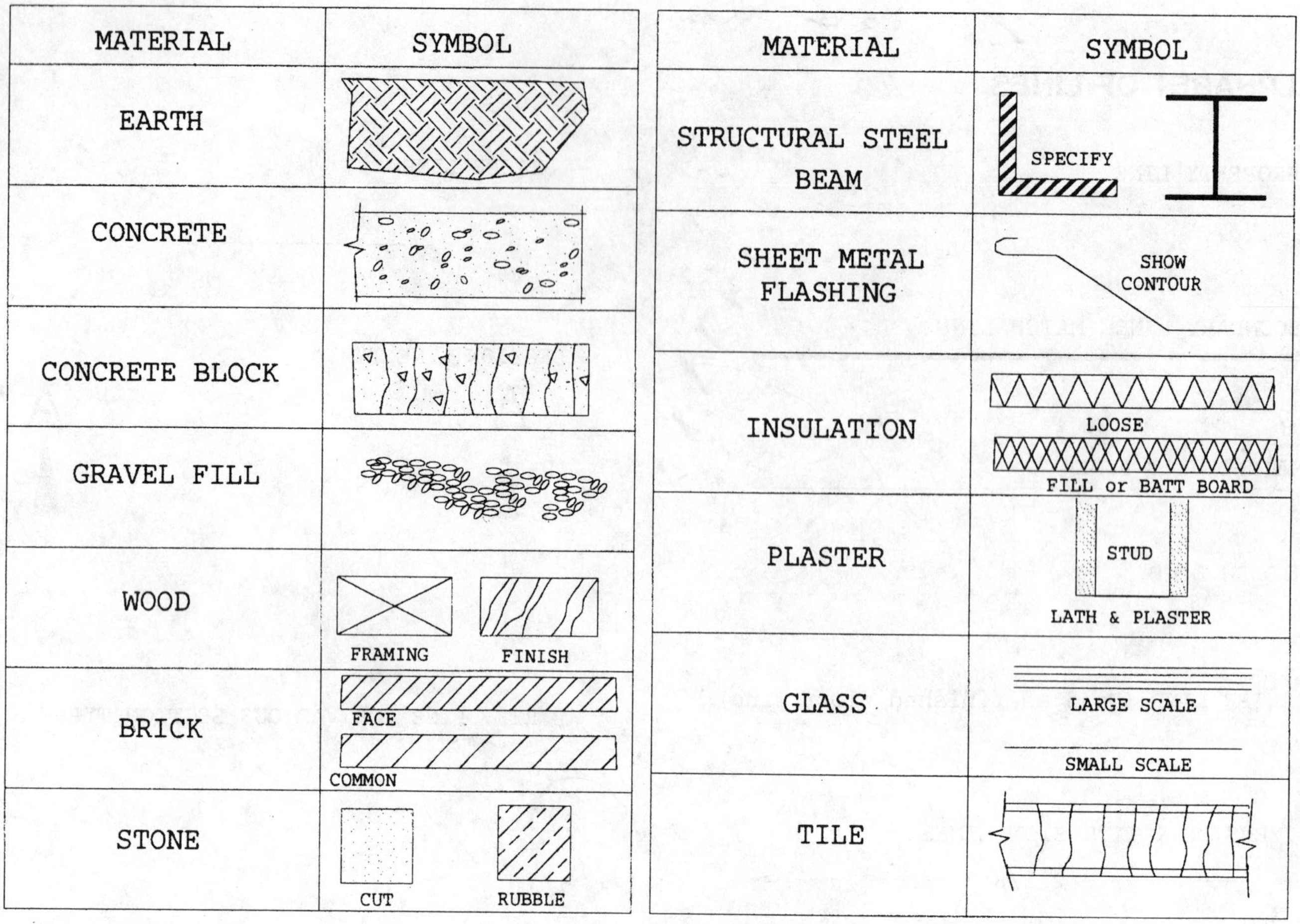

MATERIAL	SYMBOL
EARTH	
CONCRETE	
CONCRETE BLOCK	
GRAVEL FILL	
WOOD	FRAMING FINISH
BRICK	FACE COMMON
STONE	CUT RUBBLE

MATERIAL	SYMBOL
STRUCTURAL STEEL BEAM	SPECIFY
SHEET METAL FLASHING	SHOW CONTOUR
INSULATION	LOOSE FILL or BATT BOARD
PLASTER	STUD LATH & PLASTER
GLASS	LARGE SCALE SMALL SCALE
TILE	

Figure C-1. Building Material Symbols

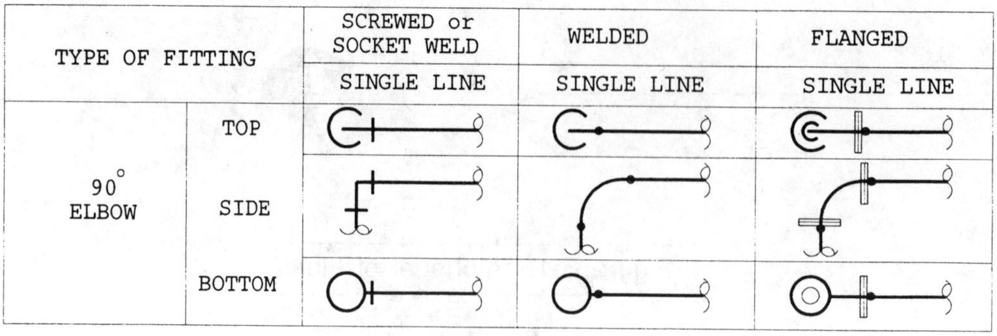

TYPE OF FITTING		SCREWED or SOCKET WELD	WELDED	FLANGED
		SINGLE LINE	SINGLE LINE	SINGLE LINE
90° ELBOW	TOP			
	SIDE			
	BOTTOM			

Figure C-2. Piping Symbols

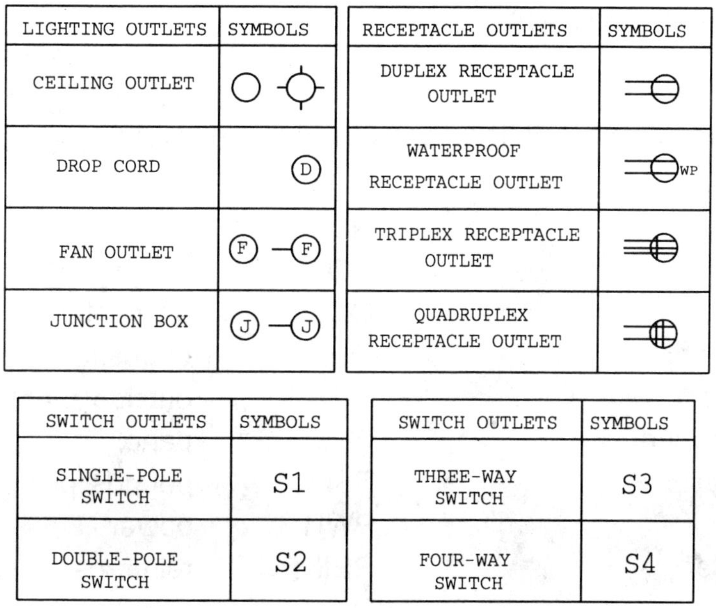

LIGHTING OUTLETS	SYMBOLS		RECEPTACLE OUTLETS	SYMBOLS
CEILING OUTLET	◯ ⬦		DUPLEX RECEPTACLE OUTLET	⊖
DROP CORD	Ⓓ		WATERPROOF RECEPTACLE OUTLET	⊖wp
FAN OUTLET	Ⓕ —Ⓕ		TRIPLEX RECEPTACLE OUTLET	⊜
JUNCTION BOX	Ⓙ —Ⓙ		QUADRUPLEX RECEPTACLE OUTLET	⊕

SWITCH OUTLETS	SYMBOLS		SWITCH OUTLETS	SYMBOLS
SINGLE-POLE SWITCH	S1		THREE-WAY SWITCH	S3
DOUBLE-POLE SWITCH	S2		FOUR-WAY SWITCH	S4

Figure C-3. Electrical Symbols

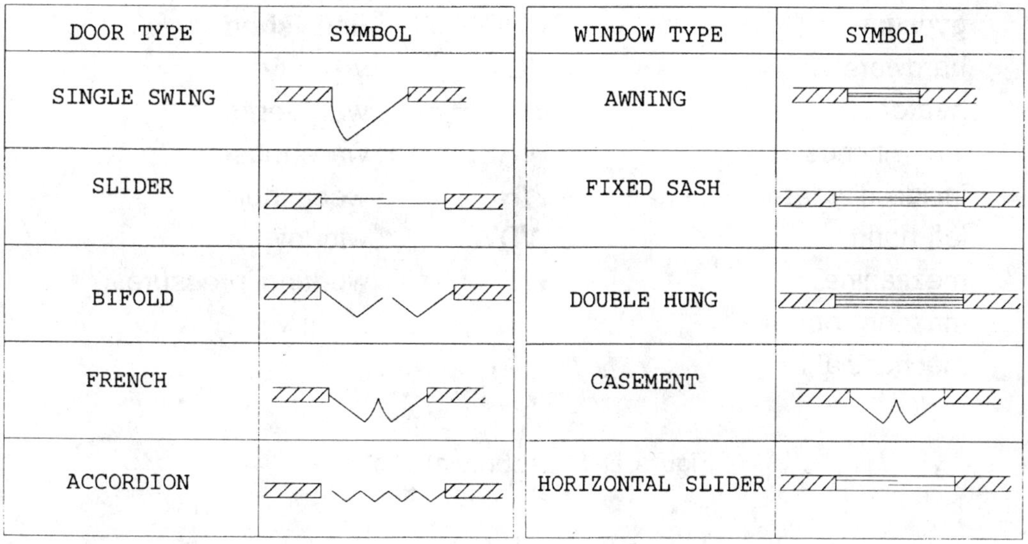

DOOR TYPE	SYMBOL		WINDOW TYPE	SYMBOL
SINGLE SWING			AWNING	
SLIDER			FIXED SASH	
BIFOLD			DOUBLE HUNG	
FRENCH			CASEMENT	
ACCORDION			HORIZONTAL SLIDER	

Figure C-4. Window and Door Symbols

APPENDIX D — ABBREVIATIONS

The following are standard abbreviations used in blueprint drawings.

ADD.	addition		N	north
AGGR	aggregate		NO.	number
L	angle		OC	on center
B	bathroom		OPP	opposite
BR	bedroom		O.D.	outside diameter
BM	bench mark		PNL	panel
BRKT	bracket		PSI	pounds per square inch
CLK	caulk		PWR	power
CHFR	chamfer		REINF	reinforce
CND	conduit		RH	right-hand
CU FT	cubic foot, feet		RFA	released for approval
DIM.	dimension		RFC	released for construction
DR	drain		RFD	released for design
DWG	drawing		RFI	released for information
ELEV	elevation		SHTHG	sheathing
ESC	escutcheon		SQ	square
FAB	fabricate		STR	structural
FLGE	flange		SYM	symbol
FLR	floor		THERMO	thermostat
GR	grade		TYP	typical
GYP	gypsum		UNFIN	unfinished
HDW	hardware		VEL	velocity
HTR	heater		WV	wall vent
" or IN.	inch, inches		WHSE	warehouse
I.D.	inside diameter		WH	weep hole
LH	left-hand		WDW	window
MEZZ	mezzanine		WP	working pressure
MO	masonry opening			
MECH	mechanical			

Figure D-1. Abbreviations

The NCCER makes every effort to keep these manuals up-to-date and free of technical errors. We appreciate your help in this process. If you have an idea for improving this manual, or if you find an error, a typographical mistake, or an inaccuracy in the *Wheels of Learning*, please write us, using this form or a photocopy. Be sure to include the exact module number, page number, a description of the problem, and the correction, if possible. We'll do our best to correct it in later editions. Thank you for your assistance.

Write: *Wheels of Learning*
National Center for Construction Education and Research
P.O. Box 141104
Gainesville, FL 32614-1104

Fax: 352-334-0932

WHEELS OF LEARNING USER UPDATE

Please let us know if you have found an inaccuracy, error, or other problem in a *Wheels of Learning* manual. Use this form or write us a letter. Please be sure to tell us the exact module name and module number, the page number, and the problem. Thanks for your help.

Craft _____ Module Name _____

Module Number _____ Page Number(s) _____

Description of Problem _____

_____

_____

(Optional) Correction of Problem _____

_____

_____

(Optional) Your Name and Address _____

_____

_____

Basic Rigging

Module 00106

NATIONAL
CENTER FOR
CONSTRUCTION
EDUCATION AND
RESEARCH

BASIC RIGGING

Objectives

Upon completion of this module, the trainee will be able to:

1. Explain and practice rigging safety.
2. Identify and explain rigging equipment.
3. Inspect rigging equipment.
4. Identify, explain, and perform crane hand signals.
5. Estimate size, weight, and center of gravity.
6. Tie knots.
7. Identify and explain types of derricks.
8. Identify and explain types of cranes.
9. Rig and move materials and equipment.

Prerequisites

Successful completion of the following Task Module(s) is required before beginning study of this Task Module: Core Task Module 00101, *Basic Safety*.

Required Student Materials

1. Personal protective safety equipment
2. Gloves
3. One medium-sized straight screwdriver
4. One 10-inch adjustable wrench

Course Map Information

This course map shows all of the *Wheels of Learning* task modules in the Core Curricula. The suggested training order begins at the bottom and proceeds up. Skill levels increase as a trainee advances on the course map. The training order may be adjusted by the local Training Program Sponsor.

Course Map: Core Curricula

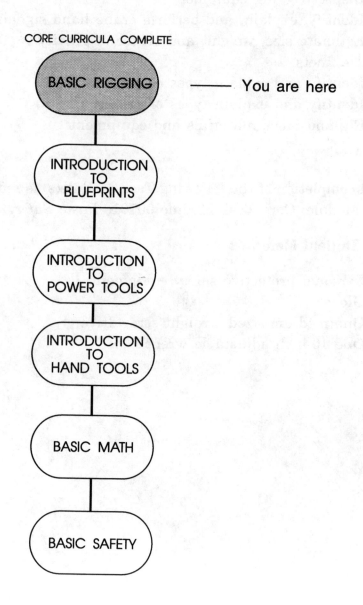

TABLE OF CONTENTS

Trade Terms Introduced in This Module.

Boom: The long, usually fabricated, part of a crane that makes it possible to position the load sheaves directly over the load to be lifted.

Bridge crane: A fabricated structural crane that operates on elevated tracks and is bridged over the lifting area.

Dunnage: Wood or other material used to support a load and keep it off the ground.

Eyebolt: A steel fitting that screws into an object. It contains an eye for attaching rigging equipment.

Gantry crane: A fabricated crane that operates on tracks at or near ground level with the traverse beam elevated and bridged over the lifting area.

Lang lay: A wire rope construction in which the wires are twisted in the same direction as the strands are twisted in the rope.

Lay of rope: A term used to describe the direction in which wires twist into strands or strands into rope.

Load (dead): The total weight of the suspended rigging.

Load (live): The weight of an object to be lifted.

Load (total): The sum of a dead load plus a live load.

Mobile: Capable of moving or being moved.

Padeye: An affixed lifting lug on a machine or piece of equipment.

Plow steel: A type of steel used in making wire rope that is tougher and stronger than the mild grades of steel.

psi: Pounds per square inch.

Regular lay: A wire rope construction in which the direction of twist of the wires is opposite that of the strands in the rope.

Sheave: A grooved pulley-wheel for changing the direction of a rope's pull.

Sling: A length of wire rope, webbing material, manila rope, or chain with an eye splice or other fitting at each end used to attach to the object to be lifted.

Strand: A group of wires twisted around a center wire, or core. Strands are twisted together to form a rope.

Traction steel: A type of steel with a strength of about 180,000 to 190,000 psi used in making wire rope.

Wedge socket: A wire rope end fitting that uses a wedge principle to hold the rope in the fitting.

Wire rope: A rope made from iron or steel wires that are formed into strands, which are in turn layed into a complete rope, sometimes called cable.

Wire rope clips: U-bolt type clips used to hold two wire ropes together.

1.0.0 INTRODUCTION

Rigging is the planned movement of an object from one place to another. It may also include moving an object from one position to another or tying down an object. For example, tools and equipment must often be lifted several floors to get them where they are needed. This moving operation must be done safely and effectively.

There are three levels of rigging knowledge: basic rigging, intermediate rigging, and advanced rigging. This basic rigging module will give you information about rigging and moving light equipment and materials. You should not assume that you are qualified to rig and move heavy equipment and materials, as this requires advanced knowledge.

2.0.0 RIGGING SAFETY

Rigging is the movement of objects such as machinery and equipment using ropes, slings, cables, rollers, hoists, and cranes. For safe rigging, the rigging tools and equipment must be in good condition and of the required strength to handle the load being rigged.

The greater the size and weight of an object, the greater the chance for injury and destruction. Estimating the weight of an object and selecting the proper rigging are the most important aspects in rigging safety. An estimate is a close calculation, not a guess. Estimates must be accurate to plus or minus 20 percent for light objects, but must be much more accurate for heavy objects.

As with any job, safety is the first consideration for rigging and moving any object. When performing rigging operations, follow these basic safety rules:

- Make sure that you and the equipment operator agree on the signals to be used before attempting to move any object.
- Thoroughly inspect all rigging equipment before using it, and cut defective rigging into unusable short pieces.
- Always wear protective gloves when handling wire rope.
- Never tie a knot in wire rope.

- Never kink a wire rope.
- Use only approved rigging equipment. Never use welded rings or field-fabricated lifting devices formed from bolts, rods, rebar, or other materials.
- Be careful to find the center of gravity of an object before attempting to lift it.
- Always test lift an object 1 to 2 inches to ensure that it is balanced.
- Never exceed the capacity of the rigging equipment.
- When using chain falls or other lifting devices, ensure that the structure the device is rigged to will support the intended load.
- Inspect rigging regularly during extended work periods.
- Never stand beneath or place any part of the body beneath a suspended load.
- Carefully follow manufacturers' instructions when fabricating wire rope and fittings or other rigging.
- Never use any rigging for anything other than its intended purpose.
- Always choose rigging that fits the object to be moved.
- Riggers must be trained to determine weights and distances and to properly select and use lifting tackle.
- Crew members must understand their specific safety responsibilities and report any unsafe conditions or practices to the proper personnel immediately.
- Personnel who work around mobile cranes must obey all warning signs and watch out for their own safety and the safety of others.
- Crew members setting up machines or handling loads must be properly trained and aware of proper machine erection and rigging procedures.
- Watch for hazards during operations, and alert the operator and signalman of dangers, such as power lines, the presence of people, other equipment, or unstable ground conditions.
- Operators must be properly trained, competent, physically fit, alert, free from the influence of alcohol, drugs, or medications, and, if required, licensed to operate mobile cranes. Good vision, judgment, coordination, and mental ability are also required. Operators who do not possess all these qualities must not be allowed to operate the equipment.
- Signalmen must have good vision and sound judgment. They must know the standard crane signals and be able to give the signals clearly. They must also have enough experience to be able to recognize hazards and signal the operator to avoid them.

3.0.0 RIGGING EQUIPMENT

Rigging equipment is available in many forms. There are special rigs for specific shaped objects, but some types of rigging can be used on many shapes. The following sections explain the common rigs and those that can be used on different shapes. The equipment covered in these sections includes the following:

- Wire rope
- Wire rope end fittings
- Wire rope slings
- Synthetic web slings
- Fiber ropes

3.1.0 WIRE ROPE

Wire rope, or cable, is made of wires, strands, and a central core. The wires of the rope are wrapped into strands, and the strands are wound around the core. *Figure 3-1* shows wire rope construction.

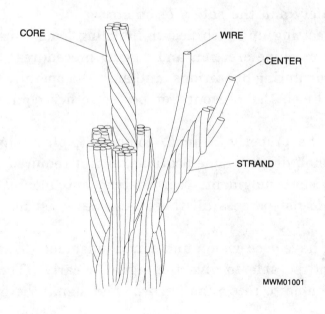

MWM01001

Figure 3-1. Wire Rope Construction

Wire rope size is identified by the number of wires in a strand and the number of strands in the rope (not counting the core). A 6 x 19 wire rope has 6 strands of 19 wires each. A 6 x 37 wire rope has 6 strands of 37 wires each. Wire ropes are made of various materials, including iron, traction steel, mild plow steel, plow steel, and improved plow steel.

Wire rope is also identified by features such as lay, preformed or nonpreformed, wire material, core material, and construction. The length of lay is the distance in which the strand makes one complete turn around the core. Wire rope is available in the following four types of lay:

- Right regular lay – The strands in the rope spiral clockwise, and the wires in the strands spiral counterclockwise.
- Left regular lay – The strands in the rope spiral counterclockwise, and the wires in the strand spiral clockwise.
- Right lang lay – Both the strands in the rope and the wires in the strands spiral clockwise.
- Left lang lay – Both the strands in the rope and the wires in the strands spiral counterclockwise.

3.2.0 WIRE ROPE END FITTINGS

Wire rope is not flexible, and you should never tie a knot in a wire rope. For these reasons, wire ropes must have end fittings that can be attached to the load. The following sections identify some basic types of wire rope end fittings:

- Eye splices
- Thimbles
- Shackles
- Sockets
- Wedge sockets
- Wire rope clips

3.2.1 Eye Splices

An eye splice in the end of a cable is the most common type of end fitting. To form an eye splice, the cable is looped, and the loose end is spliced into the cable. Then a metal sleeve is pressed around the splice to fasten the loop. *Figure 3-2* shows an eye splice with a pressed metal sleeve.

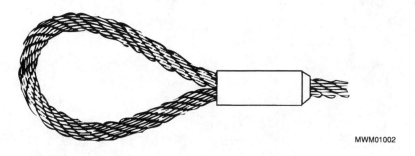

MWM01002

Figure 3-2. Eye Splice with Pressed Metal Sleeve

3.2.2 Thimbles

A thimble is a common part of many end fittings. A thimble is a grooved ring of metal that fits inside a loop of wire rope to protect the wire from wear and overbending. *Figure 3-3* shows several types of thimbles.

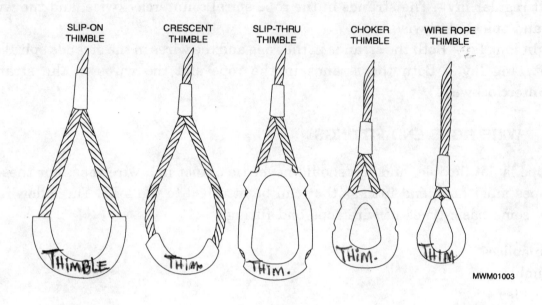

SLIP-ON
THIMBLE

CRESCENT
THIMBLE

SLIP-THRU
THIMBLE

CHOKER
THIMBLE

WIRE ROPE
THIMBLE

MWM01003

Figure 3-3. Thimbles

3.2.3 Shackles

A shackle, or clevis, is a U-shaped fitting with a removeable round pin. Some shackle pins are cotter pinned, but most have a screw pin. The pin is removed from the shackle, and the shackle is slipped through the wire rope end and attached to the object to be moved. The pin is then inserted to complete the fitting. *Figure 3-4* shows two types of shackles and their uses.

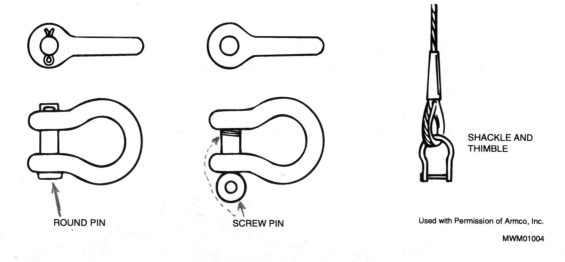

SHACKLE AND
THIMBLE

ROUND PIN

SCREW PIN

Used with Permission of Armco, Inc.

MWM01004

Figure 3-4. Shackles and Uses

3.2.4   Sockets

Sockets are wire end fittings that are permanently attached to the end of a wire rope. They provide easy attachment of other fittings. Sockets can be either open or closed type. *Figure 3-5* shows the two types of sockets.

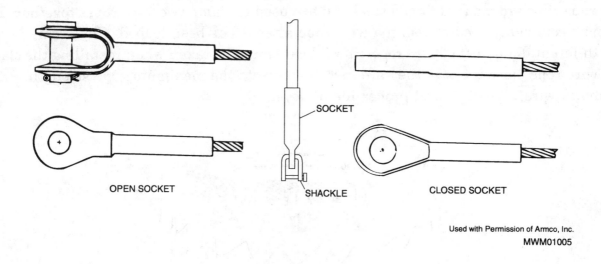

OPEN SOCKET

SOCKET

SHACKLE

CLOSED SOCKET

Used with Permission of Armco, Inc.

MWM01005

Figure 3-5.   Sockets

3.2.5 Wedge Sockets

Wedge sockets are similar to shackles. To attach a wedge socket to the cable end, the cable is looped and inserted into the socket, then the wedge is inserted to hold the cable. The cable must be threaded through the wedge properly or it will not hang beneath the center of the cable. To attach a hook to the socket, the socket pin is removed and the hook is attached, then the pin is reinserted. *Figure 3-6* shows a wedge socket fitting.

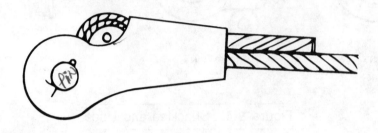

MWM01006

Figure 3-6. Wedge Socket Fitting

3.2.6 Wire Rope Clips

Wire rope clips are made of forged steel and are used to clamp two wire ropes together. They are most commonly used to clamp a wire rope after it has been looped. Clips come in sizes to fit different diameters of wire rope. Special care must be taken when installing the clamps on a wire rope since the way they are installed affects the maximum rope strength. *Figure 3-7* shows wire rope clips and proper installation.

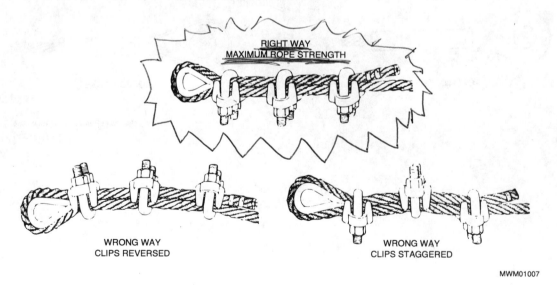

RIGHT WAY
MAXIMUM ROPE STRENGTH

WRONG WAY
CLIPS REVERSED

WRONG WAY
CLIPS STAGGERED

MWM01007

Figure 3-7. Wire Rope Clips Installation

3.3.0 WIRE ROPE SLINGS

A sling is a device that attaches a load to a crane hook. Special care should be taken when selecting a sling to ensure that it is of adequate capacity to handle the load. The size, type, and load capacity of a sling are rated by the manufacturer and are indicated on a metal ID tag attached to the sling. You should never use a sling without knowing its capacity. If the ID tag has been lost or removed from the sling, you can determine its capacity from a rigging chart. A safety margin is figured in when slings are tested, so you do not have to figure a safety margin when choosing a sling. Therefore, a 50,000-pound sling will lift 50,000 pounds. *Figure 3-8* shows a sling identification tag.

MWM01008

Figure 3-8. Sling Identification Tag

Wire rope slings must be stored properly to prevent damage. They should be stored on sling racks, with those of the same size hung together. They may also be stored in boxes or chests but should be coiled and tied.

The following three basic sling patterns are used for wire rope slings:

- Bridle slings
- Choker hitch slings
- Basket hitch slings

3.3.1 Bridle Slings

A bridle sling pattern is actually an extension of the crane cable. It has a loop or ring to attach the crane hook to a single or multiple wire rope and a hook or hooks to attach to the load.

Bridle slings are made in three basic patterns: the two-leg pattern, the three-leg pattern, and the double loop (four-leg) pattern. A bridle sling can be identified by its pear-shaped loop that attaches to a crane hook. Bridle slings are used to handle many types and shapes of objects. *Figure 3-9* shows some bridle slings and their uses.

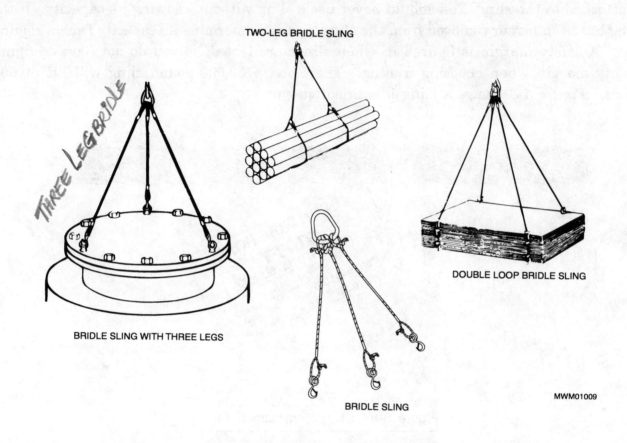

Figure 3-9. Bridle Slings and Uses

One type of two-leg bridle sling is the equalizing thimble sling. The loop that attaches to a crane hook can be positioned at any place on the sling rope. This allows the rigger to place the loop off-center to balance loads that are heavier on one end than they are on the other. *Figure 3-10* shows an equalizing bridle sling.

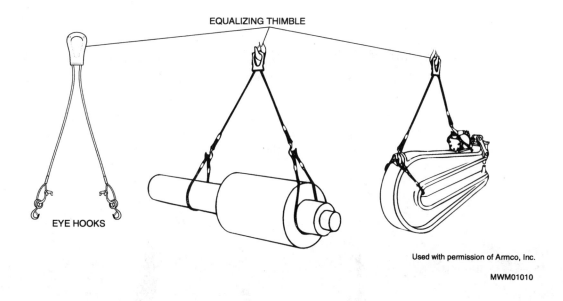

EQUALIZING THIMBLE

EYE HOOKS

Used with permission of Armco, Inc.

MWM01010

Figure 3-10. Equalizing Bridle Sling

Bridle slings are also used with pipe hooks. Pipe hooks are specialized hooks that are used primarily for lifting pipe. *Figure 3-11* shows pipe hooks used with a bridle sling.

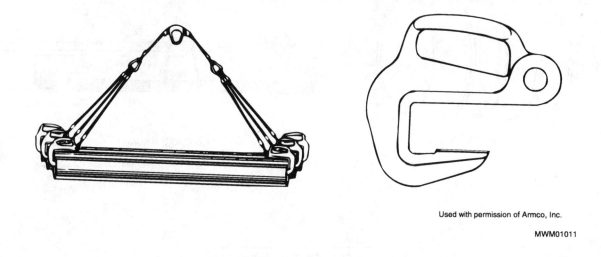

Used with permission of Armco, Inc.

MWM01011

Figure 3-11. Pipe Hooks with Bridle Sling

3.3.2 Choker Hitch Slings

Choker hitch slings are different from bridle slings in that they choke up on, or squeeze, the load. The choker hitch sling is used on objects that are hard to hook to, such as barrels, round stock, and lumber. Two types of choker hitch slings are the single strand and the dual hook chokers. *Figure 3-12* shows these two types of choker hitch slings.

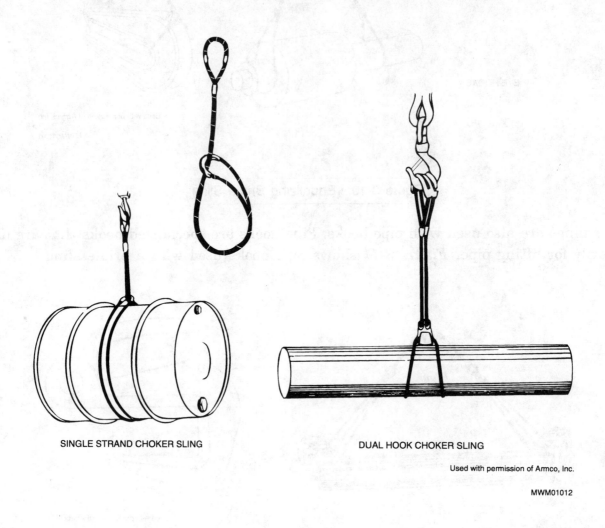

SINGLE STRAND CHOKER SLING

DUAL HOOK CHOKER SLING

Used with permission of Armco, Inc.

MWM01012

Figure 3-12.   Choker Hitch Slings

Another type of choker hitch sling is the sliding hook choker sling. The sliding hook choker sling has a metal hook that slides on the choker rope to choke up on the load. *Figure 3-13* shows a sliding hook choker sling.

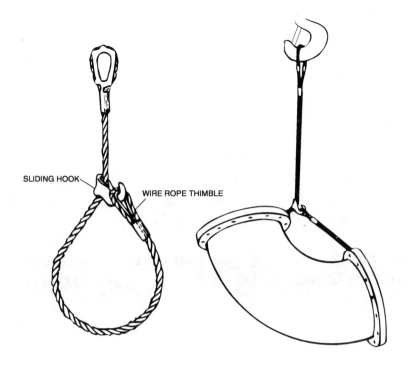

SLIDING HOOK

WIRE ROPE THIMBLE

MWM01013

Figure 3-13. Sliding Hook Choker Sling

3.3.3   Basket Hitch Slings

A basket hitch, or general purpose, sling is simply a wire rope with an eye splice in each end. This type of sling is very versatile and is used to lift many types of loads. *Figure 3-14* shows a basket hitch sling.

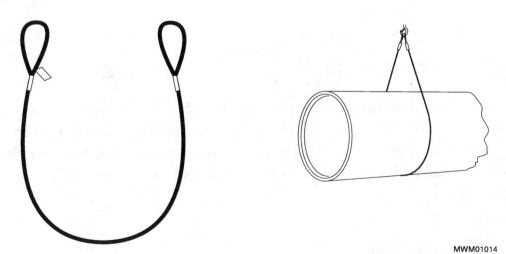

MWM01014

Figure 3-14. Basket Hitch Sling

WARNING!	A basket hitch used in this manner requires lifting with two slings, one on each end to prevent the load from slipping out of the sling.

3.4.0   SYNTHETIC WEB SLINGS

Synthetic web slings are usually made of nylon or Dacron. They are softer and more flexible than wire rope slings. Synthetic web slings are generally used to handle loads that have finished, easily damaged surfaces. Synthetic web slings should be stored in a cool, dark place, since long exposure to sunlight will damage the material.

The type number, stock number, webbing type, width, length, and capacity of the sling are recorded on an identification tag attached to each sling. *Figure 3-15* shows an identification tag.

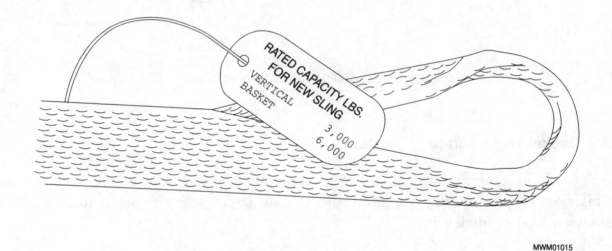

MWM01015

Figure 3-15.   Identification Tag

In a synthetic web sling, the inner yarns carry most of the load weight. Colored yarns are woven into this inner layer. If the protective outer layer of the webbing is worn away or damaged, these colors become visible. This indicates when a sling should be replaced. You should never use a sling when these colors are visible. *Figure 3-16* shows the inner yarns of a synthetic web sling.

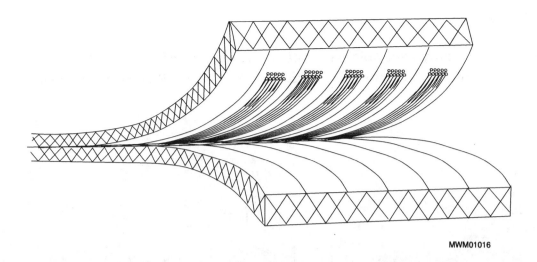

MWM01016

Figure 3-16. Inner Yarns of Synthetic Web Sling

CAUTION   Never join two slings together without the use of a shackle. Looping or tying the slings together will damage them.

The following three basic types of synthetic web slings are available:

* Choker slings
* Basket slings
* Bridle slings

3.4.1   Choker Slings

A choker sling is usually a single-leg line with slip-through triangle ends for easy handling. One metal end of the sling slips through the other metal end of the sling. This type of sling makes a smooth, even choke on a load. *Figure 3-17* shows a choker sling with triangle fittings.

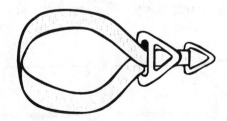

MWM01017

Figure 3-17. Choker Sling with Triangle Fittings

3.4.2   Basket Slings

Basket slings are usually available with one of three types of steel or aluminum end fittings: the hook, the triangle, or the shackle. *Figure 3-18* shows a basket sling with shackle ends.

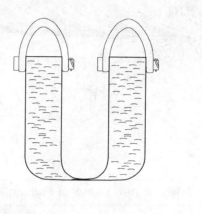

MWM01018

Figure 3-18.   Basket Sling with Shackle Ends

Basket slings are also available with one of two types of fabric ends, the flat eye and the twisted eye. On a flat eye sling, the webbing is folded back and sewn flat against the sling body. *Figure 3-19* shows a flat eye basket sling.

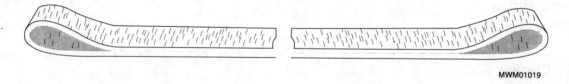

MWM01019

Figure 3-19.   Flat Eye Basket Sling

On a twisted eye sling, the fabric is twisted 180 degrees before sewing. This forms an eye that lays at 90 degrees to the sling body. *Figure 3-20* shows a twisted eye basket sling.

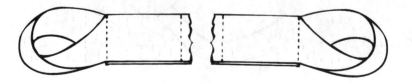

MWM01020

Figure 3-20.   Twisted Eye Basket Sling

A stable type of basket sling is the wide body sling. On a wide body sling, the eyes are folded and sewn to form hook openings. *Figure 3-21* shows two types of wide body slings.

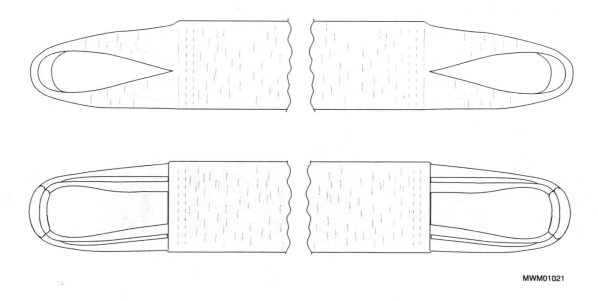

MWM01021

Figure 3-21. Wide Body Slings

Another type of basket sling is the endless grommet sling. This sling is overlapped and sewn into one large loop. Areas of hook contact may be tapered and reinforced for longer wear. *Figure 3-22* show an endless grommet sling.

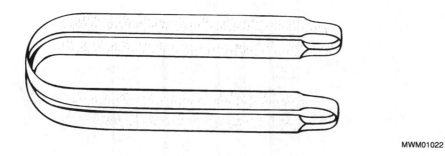

MWM01022

Figure 3-22. Endless Grommet Sling

3.4.3 Bridle Slings

Bridle slings have two or more legs. They are equipped with pear-shaped fittings to connect to a crane hook. *Figure 3-23* shows a bridle sling with two legs.

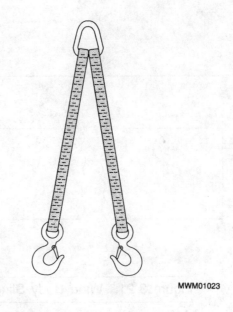

MWM01023

Figure 3-23. Bridle Sling

The leg end of a bridle sling can be equipped with a hook, triangle, shackle, or sewn eyes. *Figure 3-24* shows the types of leg fittings.

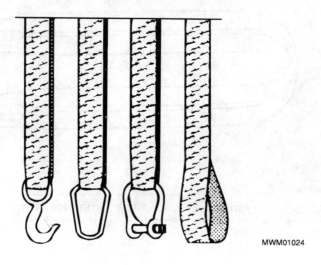

MWM01024

Figure 3-24. Leg Fittings

3.5.0 FIBER ROPES

Fiber rope is made of either plant fiber or synthetic fiber. Fiber rope usually contains three strands, but in some special cases has four, five, or six strands.

Plant fiber ropes are made of manila, jute, sisal, and hemp. Manila is the strongest of these ropes. No. 1 manila is the only plant fiber rope that is recommended for rigging. Most rope manufacturers identify No. 1 manila rope with some mark, such as colored inlaid fibers. Manila rope that has no marking is usually of a lesser grade manila and is not recommended for rigging.

Synthetic ropes are made of nylon, Dacron, polypropylene, polyester, or other special synthetic material. The different types of synthetic ropes have different strength, resistance to chemicals, resistance to abrasion, resistance to stretch, and shock-absorbing capabilities. You should follow the manufacturer's recommendations when selecting fiber rope for rigging. Like synthetic web slings, rope should be stored out of direct sunlight.

REVIEW QUESTIONS #1

1. What are the three parts of a wire rope?
2. How many strands does a 6 x 19 wire rope have?
3. Is it acceptable to tie a knot in a wire rope?
4. Which is the most common type of wire rope end fitting?
5. What is the name of a U-shaped end fitting with a removeable pin?
6. A sling is a device that attaches a load to a _____ _____.
7. How many basic patterns are bridle slings made in?
8. What is the name of a general purpose sling that has an eye splice in each end?
9. Why should synthetic web slings be stored in a dark place?
10. What is the name of a synthetic sling that has slip-through triangle ends?

PERFORMANCE/LABORATORY EXERCISE

1. Practice identifying different types of rigging equipment that is furnished in the class.

4.0.0   INSPECTING RIGGING EQUIPMENT

Rigging equipment must be properly inspected before use, and at regular intervals during extended use, to ensure safe and proper operation. Safety is the main consideration when inspecting rigging equipment. Rigging that is worn or damaged can present dangerous situations. You should check your rigging equipment regularly to ensure that it is in safe working condition. These sections explain the inspection procedures for the following equipment:

- Wire rope
- Synthetic web slings

- Hooks, shackles, and sockets
- Equipment to be rigged

4.1.0 WIRE ROPE

Wire rope is used on lifting devices, such as cranes, tuggers, and lifts, and to make wire rope slings. Wire rope is strong and versatile when it is in good condition. The strength and flexibility of wire rope can be seriously impaired if the rope is worn or damaged. The most common damage that occurs to wire ropes is broken wires. When inspecting wire rope, look for the following defects:

WARNING!   Always wear gloves when you inspect or handle new or
used wire rope to prevent possible injury to your hands.

- Broken wires – If there are six broken wires in one rope lay or three broken wires in one strand in one rope lay, the rope should be rejected.
- Corrosion – If there is evidence of deterioration from corrosion, the rope should be rejected.
- Kinking or crushing – If the rope shape has been distorted, it should be rejected.
- Evidence of heat damage – The rope should be rejected.
- Unlaying (unravelling) of a splice – The rope should be rejected.
- Core protrusion – The rope should be rejected. *Figure 4-1* shows core protrusion.
- Strand or wire slippage at an end fitting – The rope should be rejected.
- More than one broken wire in the vicinity of a fitting – The rope should be rejected.

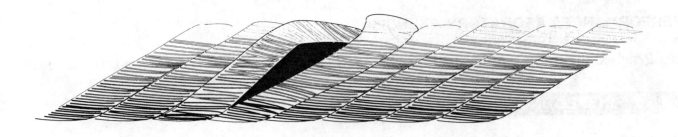

MWM01069

Figure 4-1. Core Protrusion

4.2.0 SYNTHETIC WEB SLINGS

Synthetic web slings should be inspected before use and at regular intervals during extended use. The regularity of inspection depends on the job being done and the amount of abuse that the rigging is subjected to. When inspecting synthetic web slings, look for the following defects:

- Colored inner yarn showing – Colored inner yarn showing through indicates damage from wear or over stressing. The sling should be rejected.
- Broken fibers – Broken fibers that can be seen or felt by the hand indicate damage from over stressing or abuse. The sling should be rejected.
- Cuts or tears – The sling should be rejected.
- Worn or damaged end fittings – The sling should be rejected.
- Excessively dirty – All slings will become dirty from use; however, inspect for excess dirt or grit penetrating the fibers. If the dirt may cause internal wear, the sling should be rejected.
- Evidence of heat or chemical damage – If the sling has been exposed to excessive heat, acid, or other chemicals, it should be rejected if it is not resistant to these conditions.

4.3.0 HOOKS, SHACKLES, AND SOCKETS

Hooks, shackles, and sockets should be inspected each time they are used. These items can easily become damaged during use. When inspecting hooks, shackles, sockets, and other miscellaneous fittings, look for the following defects:

- Bent or broken shackle pins – If a shackle pin becomes bent or broken, the shackle should be rejected.
- Disfiguration – If a fitting becomes bent, twisted, or otherwise damaged, it should be rejected.
- Broken or missing cotter pins – Broken or missing cotter pins should be replaced.
- Loose shackle pins – Shackle pins should be checked frequently to ensure that they are tight.
- Hook safety latch missing or damaged – Damaged or missing hook safety latches must be replaced. Never use a hook without a safety latch.
- Hook cracks – The hook should be rejected. Never use a hook that has cracks in the metal.
- Bent hook – A bent or damaged hook should be rejected.

4.4.0 EQUIPMENT TO BE RIGGED

Before rigging a load or piece of equipment, an inspection should be made to ensure that the load can be lifted safely. When inspecting a load or piece of equipment, look for the following conditions:

- Padeyes damaged – Padeyes may be damaged or defective. Inspect them carefully.
- Eyebolts not threaded all the way in – Eyebolts can sometimes come unscrewed in equipment. Make sure that they are tightened all the way and locked with a locknut. There are two types of eyebolts, straight and with a shoulder. The eyebolt that has a shoulder is suitable for angular lifts; the straight one is suitable for straight lifts only.
- Loose parts – Make sure that there are no loose parts that could fall off when the load is lifted.

REVIEW QUESTIONS #2

1. What is the most common damage that occurs to wire rope?
2. Should you use a synthetic sling that is excessively dirty?
3. How often should shackles be inspected?

PERFORMANCE/LABORATORY EXERCISE

1. Practice inspecting rigging equipment.

5.0.0   CRANE HAND SIGNALS

Cranes are noisy and are usually operated in noisy construction areas. There are also times when the crane operator cannot see the load or the area where it is to be placed. For these reasons, hand signals are used to direct the crane operator. To properly and safely direct the crane operator, some basic hand signals must be learned. You must be able to use these signals without mistakes. Radios are sometimes used to give directions to the operator when the flagman cannot be easily seen by the operator. Although hand signals are standard, the flagman and the operator must agree on the signals to be used before attempting a lift.

These sections explain the following:

- Hand signals for crawler and telescoping boom cranes
- Hand signals for tower and gantry cranes

5.1.0 HAND SIGNALS FOR CRAWLER AND TELESCOPING BOOM CRANES

The following basic hand signals must be learned for crawlers and telescoping boom type cranes:

- Raise load
- Raise load slowly
- Lower load
- Lower load slowly
- Raise boom
- Raise boom slowly
- Lower boom
- Lower boom slowly
- Raise boom, lower load
- Lower boom, raise load
- Raise boom, hold load
- Swing load
- Travel crane
- Stop
- Dog everything
- Extend boom
- Retract boom
- Extend boom (one-handed signal)
- Retract boom (one-handed signal)
- Use main hoist
- Use whip line

5.1.1 Raise Load

The raise load signal tells the operator to raise the load, using only the hook and wire. The boom should not move. To produce this signal, point your index finger skyward and move it in a circular motion. *Figure 5-1* shows the raise load signal.

MWM01025

Figure 5-1. Raise Load Signal

5.1.2 Raise Load Slowly

The raise load slowly signal tells the operator to slowly raise the load, using only the hook and wire. To produce this signal, make the raise load signal while holding the other hand flat above the signal. *Figure 5-2* shows the raise load slowly signal.

MWM01026

Figure 5-2. Raise Load Slowly Signal

5.1.3 Lower Load

The lower load signal tells the operator to lower the load, using only the hook and wire. The boom should not move. To produce this signal, point your index finger toward the ground and move it in a circular motion. *Figure 5-3* shows the lower load signal.

MWM01027

Figure 5-3. Lower Load Signal

5.1.4 Lower Load Slowly

The lower the load slowly signal tells the operator to slowly lower the load, using only the hook and wire. To produce this signal, make the lower load signal while holding your other hand flat beneath the signal. *Figure 5-4* shows the lower load slowly signal.

MWM01028

Figure 5-4. Lower Load Slowly Signal

5.1.5 Raise Boom

The raise boom signal tells the operator to raise the boom.

WARNING!	Before giving this signal, be sure that there are no obstructions, such as electric wires or building members, that the boom might contact when raised.

To produce this signal, close your fist and point your thumb skyward. *Figure 5-5* shows the raise boom signal.

MWM01029

Figure 5-5. Raise Boom Signal

5.1.6 Raise Boom Slowly

The raise boom slowly signal tells the operator to slowly raise the boom. To produce this signal, make the raise boom signal while holding the other hand flat beneath the signal. *Figure 5-6* shows the raise boom slowly signal.

MWM01030

Figure 5-6. Raise Boom Slowly Signal

5.1.7 Lower Boom

The lower boom signal tells the operator to lower the boom. To produce this signal, close your fist and point your thumb toward the ground. *Figure 5-7* shows the lower boom signal.

MWM01031

Figure 5-7. Lower Boom Signal

5.1.8   Lower Boom Slowly

The lower boom slowly signal tells the operator to slowly lower the boom. To produce this signal, make the lower boom signal and hold the other hand flat beneath the signal. *Figure 5-8* shows the lower boom slowly signal.

MWM01032

Figure 5-8.   Lower Boom Slowly Signal

5.1.9   Raise Boom, Lower Load

Sometimes the boom and load signals must be given together. The raise boom, lower load signal tells the operator to raise the boom and lower the load simultaneously. To produce this signal, make the raise boom signal with one hand and the lower load signal with the other. *Figure 5-9* shows the raise boom, lower load signal.

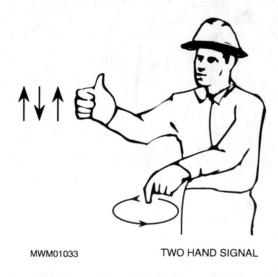

MWM01033   TWO HAND SIGNAL

Figure 5-9.   Raise Boom, Lower Load Signal

5.1.10 Lower Boom, Raise Load

The lower boom, raise load signal tells the operator to lower the boom and raise the load simultaneously. To produce this signal, make the lower boom signal with one hand and the raise load signal with the other. *Figure 5-10* shows the lower boom, raise load signal.

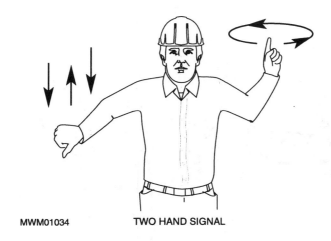

MWM01034   TWO HAND SIGNAL

Figure 5-10. Lower Boom, Raise Load Signal

5.1.11 Raise Boom, Hold Load

The raise boom, hold load signal tells the operator to raise the boom without raising or lowering the load with the hook and wire. To produce this signal, make the raise boom signal with one hand and make a clenched fist with the other. *Figure 5-11* shows the raise boom, hold load signal.

MWM01035

Figure 5-11. Raise Boom, Hold Load Signal

5.1.12 Swing Load

The swing load signal tells the operator to swing the boom and load to the left or right. To produce this signal, extend your arm and point in the direction that in which you wish the load to swing. *Figure 5-12* shows the swing load signal.

MWM01036

Figure 5-12. Swing Load Signal

5.1.13 Travel Crane

The travel crane signal tells the operator to move the entire crane to another position. To produce this signal, hold an open hand about head high and take a step in the direction you want the crane to move.

WARNING!   Be sure that there are no people or obstructions in the path of the crane.

Figure 5-13 shows the travel crane signal.

MWM01037

Figure 5-13. Travel Crane Signal

5.1.14 Stop

The stop signal tells the operator to stop the operation being performed. To produce this signal, extend both open hands straight out from your sides. *Figure 5-14* shows the stop signal.

MWM01038

Figure 5-14. Stop Signal

5.1.15 Dog Everything

The dog everything signal tells the operator to shut down all operation and hold everything in its present state. *Figure 5-15* shows the dog everything signal.

MWM01039

Figure 5-15. Dog Everything Signal

5.1.16 Extend Boom

Cranes with telescoping booms require some additional signals for extending and retracting the boom. The extend boom signal tells the operator to extend, or push out, the boom. To produce this signal, hold both hands in front of you with your fists closed and your thumbs pointing outward. *Figure 5-16* shows the extend boom signal.

MWM01040

Figure 5-16. Extend Boom Signal

CORE CURRICULA TRAINEE TASK MODULE 00106

5.1.17 Retract Boom

The retract boom signal tells the operator to retract, or pull in, the boom. To produce this signal, hold both hands in front of you with your fists closed and your thumbs pointing inward toward one another. *Figure 5-17* shows the retract boom signal.

MWM01041

Figure 5-17. Retract Boom Signal

5.1.18 Extend Boom (One-Handed Signal)

There are times when a rigger must give a signal with one hand because the other is holding a tag line. In this case, the extend boom signal must be given with one hand. To produce this signal, hold one hand with your fist closed against your chest and your thumb pointing skyward. *Figure 5-18* shows the one-handed extend boom signal.

MWM01042

Figure 5-18. Extend Boom (One-Handed Signal)

5.1.19 Retract Boom (One-Handed Signal)

To produce the extract boom signal with one hand, hold one hand with your fist closed against your chest and your thumb pointing outward. *Figure 5-19* shows the one-handed retract boom signal.

MWM01043

Figure 5-19. Retract Boom (One-Handed Signal)

5.1.20 Use Main Hoist

To produce this signal, close one fist and bump the top of your head with your fist. *Figure 5-20* shows the use main hoist signal.

MWM01070

Figure 5-20. Use Main Hoist Signal

5.1.21 Use Whip Line

To produce this signal, close one fist and bend your elbow at a 90-degree angle. Open your other hand and bump your bent elbow with that hand. *Figure 5-21* shows the use whip line signal.

MWM01071

Figure 5-21. Use Whip Line Signal

5.2.0 HAND SIGNALS FOR TOWER AND GANTRY CRANES

Flagging tower and gantry cranes involves most of the signals for crawler and telescoping cranes with a few additional signals. These sections explain the following signals used for tower and gantry cranes:

*   Travel bridge
*   Travel trolley
*   Stop
*   Emergency stop
*   Select trolley

5.2.1 Travel Bridge

The travel bridge signal tells the operator to move the bridge in one direction or the other. To produce this signal, raise an open hand to about shoulder height and make a pushing motion in the direction that you want the bridge to move. *Figure 5-22* shows the travel bridge signal.

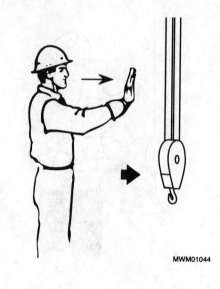

MWM01044

Figure 5-22. Travel Bridge Signal

5.2.2 Travel Trolley

The travel trolley signal tells the operator to move the trolley in one direction or the other. To produce this signal, hold one closed hand at about shoulder height with your thumb pointing in the direction that you want the trolley to move. *Figure 5-23* shows the travel trolley signal.

MWM01045

Figure 5-23. Travel Trolley Signal

5.2.3 Stop

The stop signal for tower or gantry cranes is different than that for crawler or telescoping cranes. To produce this signal, extend one arm with your palm down. *Figure 5-24* shows the stop signal.

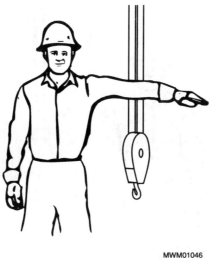

MWM01046

Figure 5-24. Stop Signal

5.2.4 Emergency Stop

In an emergency situation, the emergency stop signal is given. To produce this signal, extend an arm with your palm down and move your hand rapidly right and left. *Figure 5-25* shows the emergency stop signal.

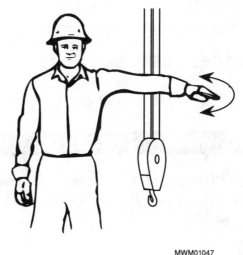

MWM01047

Figure 5-25. Emergency Stop Signal

5.2.5 Select Trolley

Some cranes have multiple trolleys. The select trolley signal tells the operator which trolley to use. To produce this signal, hold up one finger for the trolley marked "1" or two fingers for the trolley marked "2." *Figure 5-26* shows the select trolley signal.

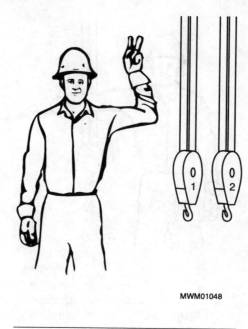

MWM01048

Figure 5-26. Select Trolley Signal

REVIEW QUESTIONS #3

1. Why are crane hand signals necessary?
2. How are directions given to the operator when the flagman cannot be easily seen?
3. Should the operator raise the boom when the raise load signal is given?
4. When you close your fist and point your thumb skyward, what signal are you giving?

PERFORMANCE/LABORATORY EXERCISE

1. Study and practice making all the hand signals listed in this section.

6.0.0   ESTIMATING SIZE, WEIGHT, AND CENTER OF GRAVITY

Before rigging an object for lifting, you must know the type of rigging and the type of lifting equipment required. To make these determinations, you must be able to perform the following activities for objects to be moved:

- Estimating size
- Estimating weight
- Finding center of gravity

6.1.0 ESTIMATING SIZE

Estimating the size of an object includes knowing the object's length, width, and/or height. The surest way to determine an object's size is to measure it using a rule or tape measure. Sometimes this cannot be done because the object is too big or hard to get to. In these cases, you must be able to estimate the object's size.

To estimate an object's size, compare the object to a measurement that you already know. It is useful to learn the measurements of different parts of your body for comparison. Most people have an arm reach, from fingertip to fingertip, of about 6 feet with the arms outstretched. You can use this measurement for a comparison to estimate the width and length of an object.

Pacing off an object is another useful way to estimate its length or width. The average person's pace, or step, is about 32 inches; so by pacing off an object and adding up the steps, you can approximate its length or width. To help you estimate the size of objects, you should know the measurements of your height, pace, and the distance from your belt to the floor. All these measurements can be used to estimate sizes. Once you know these and other sizes, practice using them to estimate sizes until you can come close to the actual size of an object.

Size estimates should be in cubic feet. Cubic feet are determined by multiplying the length times the width times the height. Once you have estimated an object's length, width, and height, you can then find an estimate of cubic feet for an object of any shape. *Figure 6-1* shows how to determine cubic feet.

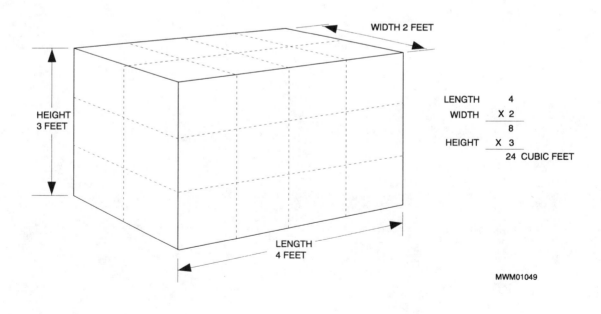

Figure 6-1. Determining Cubic Feet

6.2.0   ESTIMATING WEIGHT

Estimating the weight of an object is important in a lifting operation. You must know the weight of the object within a few hundred pounds to prevent using the wrong rigging or overloading the lifting equipment. This could cause serious accidents, damages, or personal injury.

Sometimes the weight and size of an object is printed on a data plate on the object or the crate it comes in. This is not always true, however, so you should be able to estimate the weight of equipment, crates, boxes, bundles, and odd-shaped objects. Estimating is a close calculation, not a guess.

It is difficult to estimate the weight of objects like pipe because pipe is hollow and has different wall thicknesses. For certain kinds of pipe, the weights can be found in charts or tables. There are also tables that list the weights of flat materials. *Table 6-1* lists standard pipe data. *Table 6-2* lists the weights of copper tubing.

Table 6-1. Standard Pipe Data

Nominal Pipe Diameter, Inches		Actual Inside Diameter, Inches	Actual Outside Diameter, Inches	Inside Area, Square Inches	Weight Per Foot, Pounds	Gallons Per Linear Foot
1/8	1/4	.269	.405	.057	.244	.0030
3/8	1/2	.364	.540	.104	.424	.0054
		.493	.675	.191	.567	.0099
		.622	.840	.304	.850	.0158
3/4	1	.824	1.050	.533	1.130	.0277
1-1/4		1.049	1.315	.864	1.678	.0449
1-1/2		1.380	1.660	1.496	2.272	.0777
		1.610	1.900	2.036	2.717	.1058
2	2-1/2	2.067	2.375	3.356	3.652	.1743
3	3-1/2	2.469	2.875	4.788	5.793	.2487
		3.068	3.500	7.393	7.575	.3840
		3.548	4.000	9.887	9.109	.5136
4	4-1/2	4.026	4.500	12.730	10.790	.6613
5	6	4.560	5.000	15.947	12.538	.8284
		5.047	5.563	20.006	14.617	1.0393
		6.065	6.625	28.890	18.974	1.5008
8	10	7.981	8.625	50.027	28.544	2.5988
12		10.020	10.750	78.854	40.483	4.0963
14		12.000	12.750	113.098	48.995	5.8752
		13.250	14.000	137.886	53.941	7.1629
16		15.250	16.000	182.650	61.746	9.4883
18		17.250	18.000	233.710	69.753	12.1407
20		19.250	20.000	291.039	77.619	15.1189
24		23.250	24.000	424.560	93.509	22.0050
26		25.250	26.000	500.742	101.435	26.0127
30		29.250	30.000	671.959	117.267	34.9069
36		35.250	36.000	975.909	141.017	50.6965

Table 6-2. Copper Tubing Weight

Nominal Size (Inches)		OD All Sizes	Type K Lb./Ft.	Type L Lb./Ft.	Type M Lb./Ft.	DWV Lb./Ft.
1/4		.375	.145	.126	N.A.	Not available in these sizes
3/8	1/2	.500	.269	.198	.145	
5/8	3/4	.625	.344	.285	.204	N.A.
		.750	.418	.362	N.A.	
		.875	.641	.455	.328	
1	1-1/4	1.125	.839	.655	.465	N.A.
1-1/2	2	1.375	1.04	.884	.682	6.50
		1.625	1.36	1.14	.940	.809
		2.125	2.06	1.75	1.46	1.07
2-1/2	3	2.625	2.93	2.48	2.03	N.A.
3-1/2		3.125	4.00	3.33	2.68	1.69
4		3.625	5.12	4.29	3.58	N.A.
		4.125	6.51	5.38	4.66	2.87

These two tables list the weights in pounds per foot. After finding the weight per foot, multiply it by the number of feet of pipe to be moved. This will give you the weight of the load.

For example, you are moving 10 lengths of standard 2-inch steel pipe, and each length is 22 feet. This means that you are moving 220 feet of pipe (22 x 10 = 220).

As shown in Table 6-1, standard 2-inch steel pipe weighs 3.65 pounds per foot. To find the load weight, multiply 220 feet of pipe times 3.65 pounds per foot. The load weight is then 803 pounds (220 x 3.65 = 803.00).

If the weight is not marked on the object or on the crate, you may be able to find the weight of a material by using a chart of weights per square foot. Multiply this weight by the number of cubic feet in the load. This will give you the weight of the load. *Table 6-3* lists the weights of flat metals per square foot. *Table 6-4* lists weights of common materials.

Table 6-3. Weights of Metal per Square Foot

Thickness in Inches	Wrought Iron	Cast Iron	Steel	Copper	Tin	Zinc	Brass	Lead
1/16	2.50	2.34	2.55	2.89	2.41	2.28	2.63	3.7
1/8	5.00	4.69	5.10	5.79	4.81	4.55	5.26	7.4
3/16	7.50	7.03	7.65	8.68	7.22	6.83	7.89	11.1
1/4	10.00	9.38	10.2	11.6	9.63	9.10	10.5	14.8
5/16	12.50	11.7	12.8	14.5	12.0	11.4	13.2	18.5
3/8	15.00	14.1	15.3	17.4	14.4	13.7	15.8	22.2
7/16	17.50	16.4	17.9	20.3	16.8	15.9	18.4	25.9
1/2	20.00	18.7	20.4	23.2	19.3	18.2	21.1	29.7
9/16	22.50	21.1	23.0	26.0	21.7	20.5	23.7	23.4
5/8	25.00	23.5	25.5	28.9	24.1	22.8	26.3	37.1
11/16	27.50	25.8	28.1	31.8	26.5	25.0	28.9	40.8
3/4	30.00	28.1	30.6	34.7	28.9	27.3	31.6	44.4
13/16	32.50	30.5	33.2	37.6	31.3	29.6	34.2	48.2
7/8	35.00	32.8	35.7	40.5	33.7	31.9	36.8	51.9
15/16	37.50	35.2	38.3	43.4	36.1	34.1	39.5	55.6
1	40.00	37.5	40.8	46.3	38.5	36.4	42.1	59.3

Table 6-4. Weights of Common Materials

Metal	Weight lbs./cu. ft.		Name of Material	Weight lbs./cu. ft.
Aluminum				160
Antimony				150
Bismuth Brass,	166	418	Bluestone Brick,	125
cast Brass, rolled	613	504	pressed Brick,	100-120
Copper, cast	523	550	common Cement,	70-90
Copper, rolled	555	1204	Portland (packed) Cement,	80-100
Gold, 24-carat	450		Portland (loose) Cement,	55-75
Iron, cast Iron,	480	712	slag (packed) Cement, slag	156
wrought Lead,	846		(loose) Chalk	15-34
commercial	655	490	Charcoal	110
Mercury, 60 deg. F	458	437	Cinder concrete	120-150
Silver Steel			Clay, ordinary	93.5
Tin, cast			Coal, hard solid	54
Zinc			Coal, hard, broken	84
			Coal, soft, solid	54
			Coal, soft, broken	23-32
Wood	**Weight lbs./cu. ft.**		Coke, loose Concrete	140-155
			or stone Earth, rammed	90-100
Ash Beech			Granite	165-170
Birch			Gravel Lime,	117-125
Cedar	35		quick (ground loose)	53
Cherry	37	40	Limestone Marble	170
Chestnut	22	30	Plaster of Paris (cast)	164
Cork	26	15	Sand	80
Cypress	27		Sandstone Shale	90-106
Ebony	71		Slate Terra	151
Elm	30	22	cotta Trap rock	162
Fir, Balsam	31			160-180
Hemlock				110
				170

On loads whose weight is unknown, you must be able to estimate the weight to within a few hundred pounds before attempting a lift. On very heavy loads, the estimate should be even more accurate. If you do not know the weight of the load and there is no chart to estimate the load, you must contact the project engineer, your foreman, or the vendor who supplied the equipment to determine the weight. Never attempt to lift a load without knowing its weight.

6.3.0　FINDING CENTER OF GRAVITY

The center of gravity of an object is the exact point at which the object is balanced in all directions: width, length, and height. It is important to find the center of gravity of a load before attempting to lift it. If the load is not balanced when it is lifted, it may twist, turn, or even fall from the rigging, presenting a dangerous situation.

To make a level lift, it is necessary to have the crane hook directly above the load's center of gravity and to have slings of the proper length to balance the load. Some equipment comes with lifting lugs, or padeyes, attached in the proper places to establish the center of gravity. On other loads, you must determine what length of rigging to use and where it should be connected to the load.

Many times, the center of gravity is the exact center of the object. On some unbalanced objects, the center of gravity is off-center. Skill in estimating the center of gravity of odd-shaped, unbalanced objects comes with experience.

Once you have estimated the center of gravity and attached the rigging, test lift the object 1 to 2 inches to ensure that it is balanced and lifting level. If the load is not balanced, lower it and readjust the rigging until the load lifts properly. *Figure 6-2* shows centered and uncentered loads.

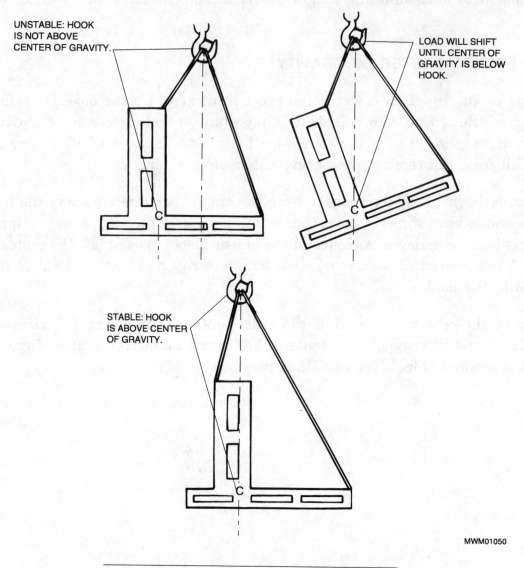

Figure 6-2. Centered and Uncentered Loads

REVIEW QUESTIONS #4

1. What is the surest way to determine an object's size?
2. How many inches is an average person's step?
3. Estimating is not a guess but a _____ _____.

PERFORMANCE/LABORATORY EXERCISE

1. Practice estimating the center of gravity of objects.

Fiber rope slings are used for lightweight lifting operations that do not require wire rope slings. Many times, it is faster to attach a fiber rope with a knot than to attach a wire rope. To use fiber rope slings, you must be able to tie a sturdy, secure knot. The following basic knots are used to attach fiber rope:

- Bowline
- Running bowline
- Timber hitch
- Half hitch
- Square
- Clove hitch
- Barrel hitch

7.1.0   BOWLINE

The bowline is the most frequently used knot in rigging. Tied properly, this knot will not slip and tighten on the load. *Figure 7-1* shows tying a bowline knot.

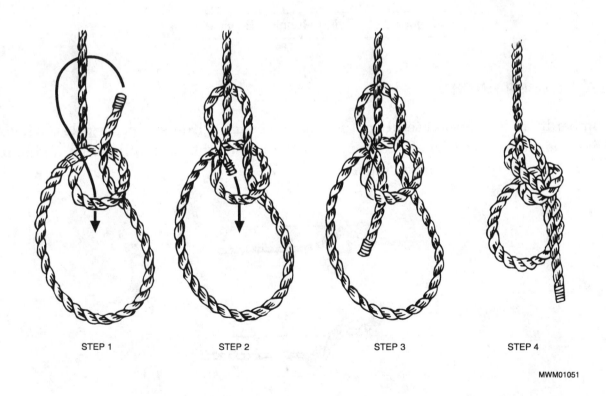

STEP 1   STEP 2   STEP 3   STEP 4

MWM01051

Figure 7-1. Tying Bowline Knot

7.2.0 RUNNING BOWLINE

The running bowline is tied like the bowline except that after the bowline is tied, the long end of the rope is slipped back through the loop made by the bowline. This makes the knot a slipknot hitch. *Figure 7-2* shows tying a running bowline knot.

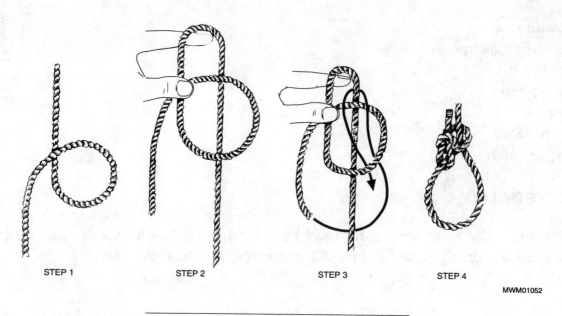

STEP 1   STEP 2   STEP 3   STEP 4

MWM01052

Figure 7-2. Tying Running Bowline Knot

7.3.0 TIMBER HITCH

The timber hitch is a quick, useful knot for rigging beams, girders, pipe, or poles. When this knot is used, there must be a steady pull on the line or the rope will loosen and slip off. *Figure 7-3* shows tying a timber hitch knot.

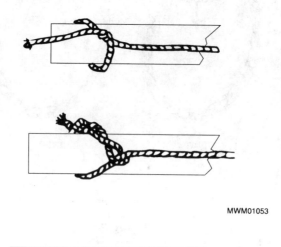

MWM01053

Figure 7-3. Tying Timber Hitch Knot

7.4.0 HALF HITCH

The half hitch is used in conjunction with the timber hitch to keep the ends of pipe or timber from revolving and to keep small diameter tanks upright and in line with the line of pull. *Figure 7-4* shows a half hitch used with a timber hitch.

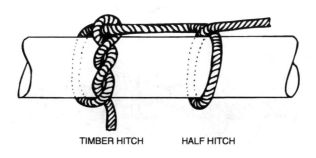

TIMBER HITCH   HALF HITCH

MWM01054

Figure 7-4. Half Hitch with Timber Hitch

7.5.0 SQUARE

A square knot is used to tie two ropes together or to tie the ends of one rope together. *Figure 7-5* shows tying a square knot.

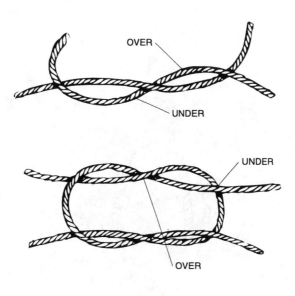

OVER

UNDER

UNDER

OVER

MWM01055

Figure 7-5. Tying Square Knot

7.6.0   CLOVE HITCH

The clove hitch is used for hoisting pipe or small diameter tanks. It will take a great deal of strain in either direction without slackening, and it is easy to untie. *Figure 7-6* shows tying a clove hitch.

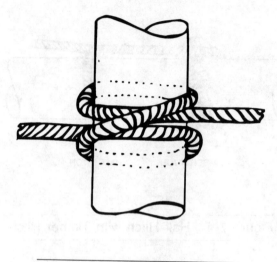

MWM01056

Figure 7-6.   Tying Clove Hitch

7.7.0   BARREL HITCH

The barrel hitch is used to hoist a barrel or a bag of fittings, concrete, sand, or gravel. *Figure 7-7* shows tying a barrel hitch.

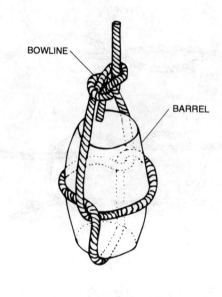

BOWLINE

BARREL

MWM01057

Figure 7-7.   Tying Barrel Hitch

REVIEW QUESTIONS #5

1. Which is the most frequently used fiber rope knot?
2. Which knot requires a steady pull on the rope to prevent it from slipping off?
3. Which knot would you use to lift a bag of fittings?

PERFORMANCE/LABORATORY EXERCISE

1. Practice tying all the knots listed in this section.

8.0.0 TYPES OF DERRICKS

A derrick is a stationary crane that is used to lift and move heavy objects. Some derricks are used primarily for lifting objects, while others provide for some movement of the load. A derrick usually carries working tackle at its upper end. It may or may not include a boom. These sections explain the following five common derricks:

- A-frame derricks
- Gin pole derricks
- Guyed derricks
- Stiff leg derricks
- Chicago boom derricks

8.1.0 A-FRAME DERRICKS

The A-frame derrick is supported by an A-frame and a guy line. The A-frame support does not move or shift and allows for only a small amount of horizontal movement. Its main job is lifting. Small A-frame derricks are manually operated. Larger A-frame derricks use electric or pneumatic power for lifting loads. An A-frame derrick consists of the following parts:

- A-frame – Supports the boom and the load.
- Boom – Supports the working tackle.
- Guy line – Holds the A-frame in an upright position and supports most of the load.
- Lead line – Is pulled by the operator to lift the load.
- Topping tackle – Raises or lowers the boom.
- Working tackle – Raises or lowers the load.

Figure 8-1 shows an A-frame derrick.

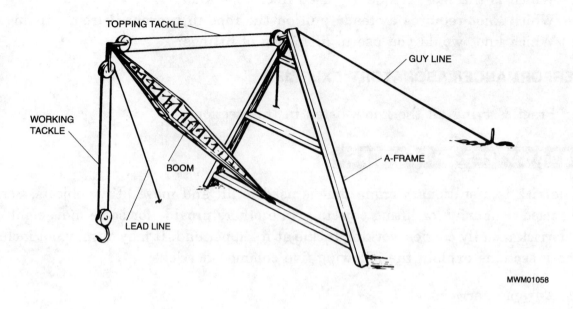

Figure 8-1. A-Frame Derrick

8.2.0   GIN POLE DERRICKS

The gin pole derrick has a single pole secured at the base to restrict horizontal movement but to allow rotation. The pole is held up by guy wires. The gin pole is mainly used for raising and lowering loads. The guy wire supports allow only a small amount of horizontal movement. Gin pole derricks are sometimes used in pairs for very heavy loads. A gin pole derrick consists of the following parts:

• Gin pole – Supports the working tackle.
• Guy wires – Support the gin pole.
• Working tackle – Raises or lowers the load.

Figure 8-2 show a gin pole derrick.

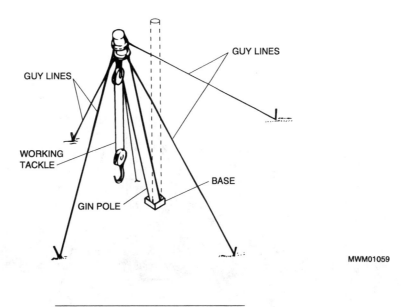

GUY LINES

GUY LINES

WORKING
TACKLE

GIN POLE

BASE

MWM01059

Figure 8-2. Gin Pole Derrick

8.3.0 GUYED DERRICKS

A guyed derrick combines the A-frame and the gin pole derricks. The guyed derrick uses a mast supported by guy wires and a rotating base that allows full-circle rotation. Heavier loads can be lifted with a guyed derrick. A guyed derrick consists of the following parts:

- Boom – Supports the working tackle.
- Collar – A rotating device at the top of the mast, also called a spider.
- Guy wires – Support the mast in an upright position and carry the load.
- Mast – Carries the tackle and supports the load.
- Rotating base – Supports the mast and boom and allows them to rotate.
- Topping tackle – Raises or lowers the boom.
- Working tackle – Raises or lowers the load.

Figure 8-3 shows a guyed derrick.

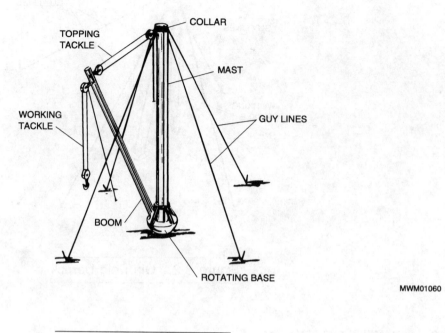

Figure 8-3. Guyed Derrick

8.4.0 STIFF LEG DERRICKS

The stiff leg derrick is similar to the guyed derrick. The mast is supported by two or more inclined struts called stiff legs connected to the top of the mast. The struts are spread 80 to 90 degrees apart to provide support in two directions. The mast and boom can swing through an arc of 270 degrees. A stiff leg derrick consists of the following parts:

* Boom – Supports the working tackle.
* Boom lines – Raise and lower the boom.
* Mast – Carries the tackle and supports the load.
* Rotating base – Supports the mast and boom and allows them to rotate.
* Bull wheel – Rotates the mast.
* Stiff legs – Support the mast in an upright position and carry the load. They are connected to the top of the mast.

Figure 8-4 shows a stiff leg derrick.

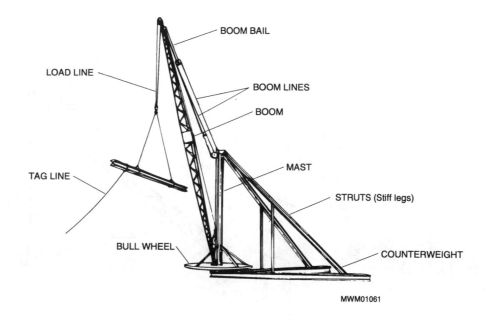

Figure 8-4. Stiff Leg Derrick

8.5.0 CHICAGO BOOM DERRICKS

The Chicago boom derrick is used to hoist equipment and supplies for high-rise construction. The derrick is usually installed by mounting it on a building column. A Chicago boom derrick consists of the following parts:

- Boom – Supports the working tackle.
- Boom tackle – Raises and lowers the boom.
- Boom base – Supports the boom and allows it to turn.
- Boom tackle base – Supports the boom tackle.

Figure 8-5 shows a Chicago boom derrick.

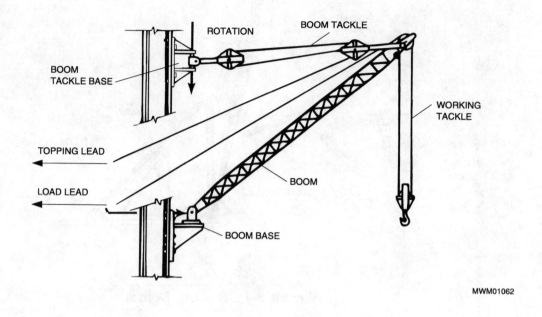

Figure 8-5. Chicago Boom Derrick

REVIEW QUESTIONS #6

1. Which type of derrick is supported by an A-frame and a guy line?
2. Which type of derrick has a single pole secured at its base?
3. Which type of derrick combines the A-frame and gin pole derricks?
4. Which type of derrick is supported by two or more stiff legs?

9.0.0   TYPES OF CRANES

Cranes are used to lift and move materials and equipment from one place to another. Some cranes are mobile, while others are stationary. These sections explain the following common types of cranes:

- Crawler cranes
- Truck-mounted cranes
- Truck-mounted hydraulic cranes
- Gantry-mounted cranes
- Tower-mounted cranes
- Hammerhead cranes
- Cherry pickers
- Drotts

CORE CURRICULA TRAINEE TASK MODULE 00106

9.1.0 CRAWLER CRANES

The crawler is the most useful crane on soft ground. The large crawler tracks make the crane both mobile and stable. The crawler crane does not have outriggers. A crawler crane consists of the following parts:

- Back stops – Telescoping pipe assembly used to stop the boom at or near the upright position.
- Boom – The long framework that supports the lifting tackle.
- Boom base – The lower part of the boom that attaches to the crane cab.
- Boom pendant lines – A block and sheave assembly used to raise or lower the boom.
- Boom point – The outward end of the top section of the boom.
- Counterweight – Mounted on the back of the crane cab; used to offset the weight of the load.
- Gantry – A frame that supports the boom harness.
- Lifting tackle – The block and sheave assembly used to raise and lower the load.

Figure 9-1 shows a crawler crane.

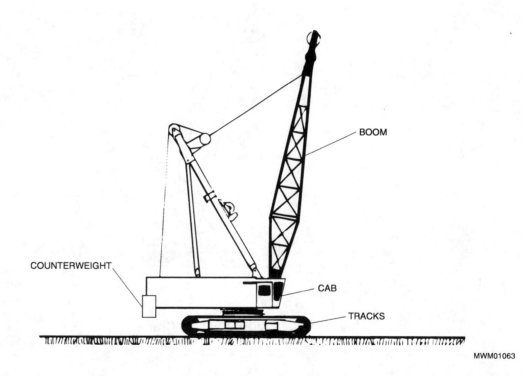

MWM01063

Figure 9-1. Crawler Crane

9.2.0 TRUCK-MOUNTED CRANES

The truck-mounted crane is the most mobile of all cranes. It can quickly and easily be moved from one job to another. The truck-mounted crane can be moved on the highway when the boom is in the transport position. *Figure 9-2* shows a truck-mounted crane and its parts.

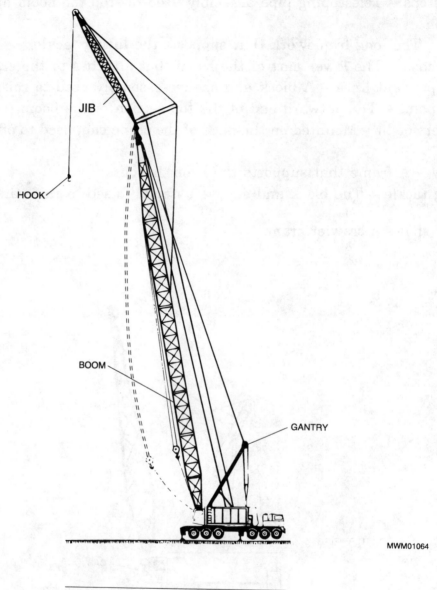

MWM01064

Figure 9-2. Truck-Mounted Crane

9.3.0 TRUCK-MOUNTED HYDRAULIC CRANES

A truck-mounted hydraulic crane has a telescoping boom that can be extended to reach a load or retracted when the crane is moved. The truck-mounted hydraulic crane uses fluid pressure to raise, lower, extend, and retract the boom. *Figure 9-3* shows a truck-mounted hydraulic crane.

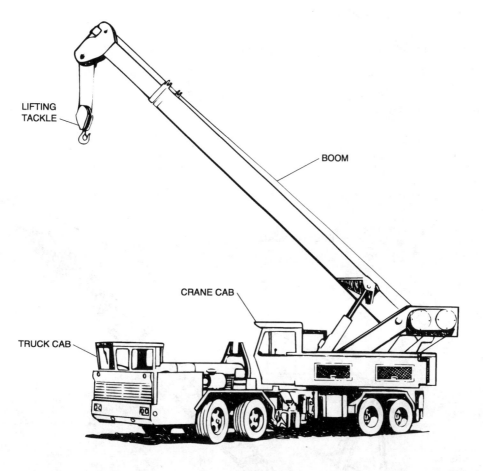

LIFTING TACKLE

BOOM

CRANE CAB

TRUCK CAB

MWM01065

Figure 9-3. Truck-Mounted Hydraulic Crane

9.4.0 GANTRY-MOUNTED CRANES

Gantry-mounted cranes are used on construction sites and on docks for loading and unloading ships. This crane is mounted on a steel frame, or gantry, and can move in a full circle. The gantry-mounted crane can also be mounted on tracks, allowing more movement. *Figure 9-4* shows a gantry-mounted crane.

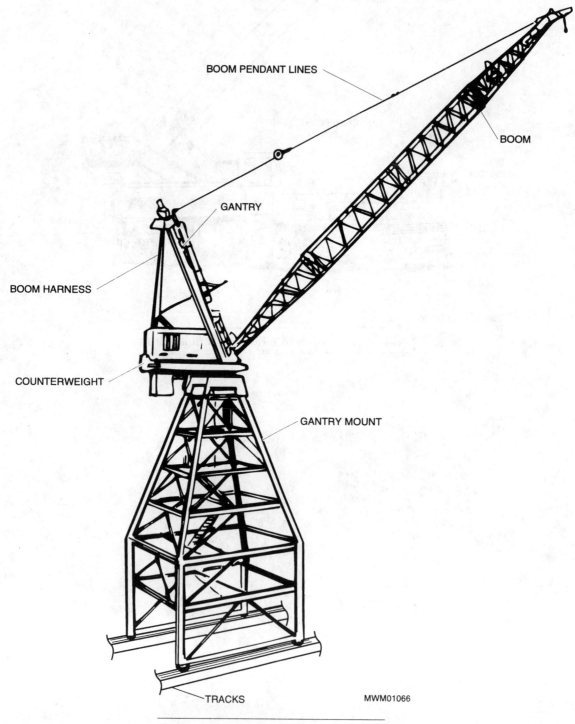

BOOM PENDANT LINES

BOOM

GANTRY

BOOM HARNESS

COUNTERWEIGHT

GANTRY MOUNT

TRACKS

MWM01066

Figure 9-4. Gantry-Mounted Crane

9.5.0 TOWER-MOUNTED CRANES

Tower-mounted cranes are used in high-rise construction. This type of crane is stationary and cannot be moved. The tower-mounted crane is usually erected at the beginning of the job and left for the duration of the job. *Figure 9-5* shows a tower-mounted crane.

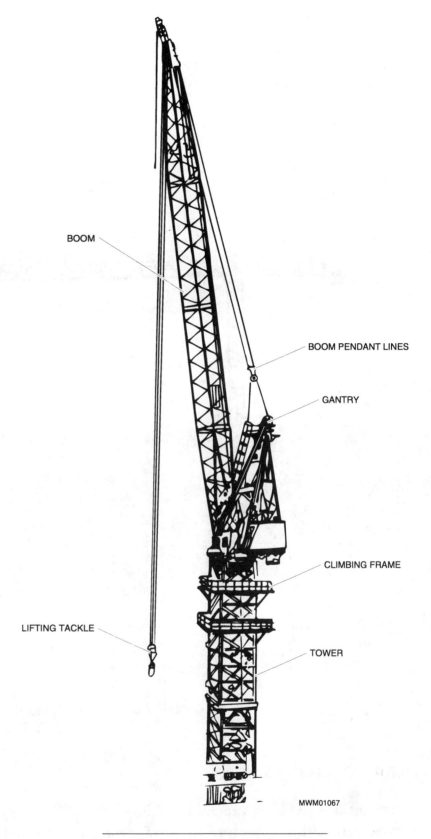

BOOM

BOOM PENDANT LINES

GANTRY

CLIMBING FRAME

LIFTING TACKLE

TOWER

MWM01067

Figure 9-5. Tower-Mounted Crane

9.6.0 HAMMERHEAD CRANES

The hammerhead crane is a type of tower crane. This crane has a long trestle that rotates around the tower on a pivot and ring. The trestle is supported by cables attached to a mast and balanced by counterweights and the machinery house on the opposite side of the mast. The crane is operated from the cab located above the pivot and ring. *Figure 9-6* shows a hammerhead crane.

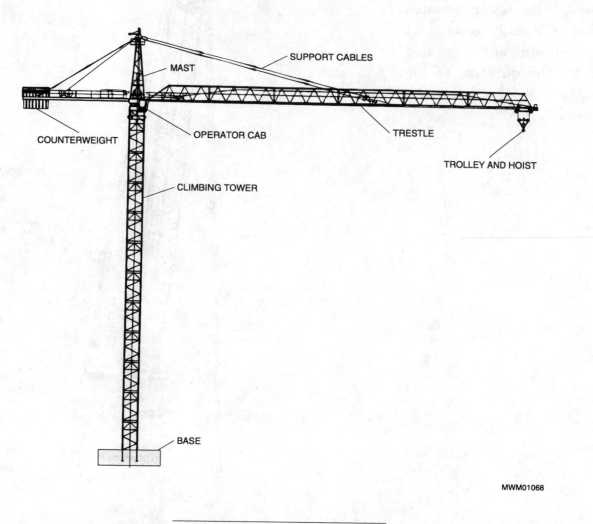

MWM01068

Figure 9-6.   Hammerhead Crane

9.7.0 CHERRY PICKERS

Cherry pickers are rough terrain mobile cranes. They have oversized tires that allow them to move across the rough terrain of construction sites and other broken ground. Two types of cherry pickers are the fixed cab and the rotating cab. *Figure 9-7* shows two types of cherry pickers.

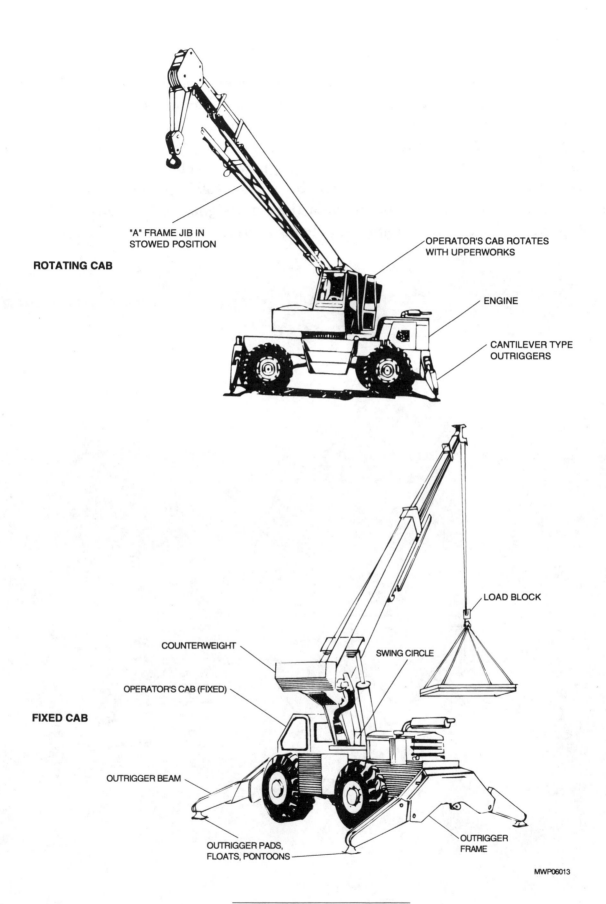

ROTATING CAB

"A" FRAME JIB IN
STOWED POSITION

OPERATOR'S CAB ROTATES
WITH UPPERWORKS

ENGINE

CANTILEVER TYPE
OUTRIGGERS

LOAD BLOCK

COUNTERWEIGHT

SWING CIRCLE

OPERATOR'S CAB (FIXED)

FIXED CAB

OUTRIGGER BEAM

OUTRIGGER PADS,
FLOATS, PONTOONS

OUTRIGGER
FRAME

MWP06013

Figure 9-7.   Cherry Pickers

9.8.0 DROTTS

A drott is a small hydraulic mobile crane. Drotts are primarily used in industrial applications where working surfaces are significantly better than those found on most construction sites. The compact size of the drott, 15 feet long by 7 feet wide, allows it to be maneuvered easily about the job site. It offers up to 8-1/2 tons of lifting capacity with the outriggers in place and up to 7-1/2 tons for lifting- and transporting-type work. Most drotts come with water-cooled diesel engines, but they are also available with gasoline-powered and propane (LP) engines. The drott also has a transport area that is designed to lock down and carry up to 15,000 pounds of materials. *Figure 9-8* shows a drott.

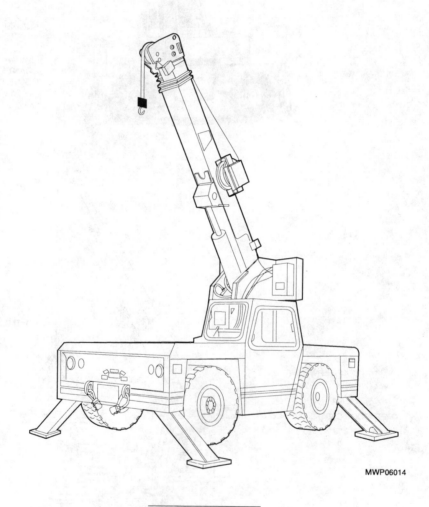

MWP06014

Figure 9-8. Drott

REVIEW QUESTIONS #7

1. Which crane is most useful on soft ground?
2. Which crane can be moved on the highway?
3. Which truck-mounted crane has a telescoping boom?
4. Which crane is sometimes mounted on tracks for movement?
5. Which type of crane is usually erected at the beginning of the job and left for the duration of the job?
6. Which crane has a long tressel that rotates a tower on a pivoted ring?
7. Which crane has a transport area for carrying materials?

10.0.0 RIGGING AND MOVING MATERIALS AND EQUIPMENT

Safely lifting and moving machinery and equipment is not automatic. It involves making several decisions regarding the object to be moved, the lifting equipment to be used, and the rigging equipment that will be used. Safely lifting and moving an object is the responsibility of the rigger. Follow these steps to rig and move materials and equipment.

Step 1   Put on safety glasses, a hard hat, and gloves.

Step 2   Determine the size of the object to be moved.

Step 3   Determine the weight of the object to be moved.

Step 4   Determine where the object will be moved to and the route that will be taken.

Step 5   Check with the crane operator when moving very heavy objects to ensure that the crane is of adequate capacity to make the lift.

Note   Everything that is suspended by the boom is considered the load. This includes crane load blocks, load lines, and all lifting tackle.

Step 6   Determine the working radius of the crane, and have the crane set up in a suitable location to make the move.

WARNING!   Ensure that the crane is parked on a suitable surface, that the outriggers are set, and that the area around the crane is roped off to prevent anyone from getting near the crane while it is in operation.

Step 7   Select the proper size, type, and capacity of rigging materials, such as shackles, wire ropes, slings, and fittings, to perform the lift.

WARNING!   Ensure that the rigging is of adequate capacity for the load and the proper configuration for the lift. Using improper rigging can result in serious or fatal injury.

Step 8 Determine the load center of gravity.

Note On some objects, the center of gravity can be found by measuring to find the center of the object. On other objects, the center of gravity can only be estimated until the load can be test lifted.

Step 9 Determine how the object will be rigged.

CAUTION The rigging must be arranged so that all parts of the load will be supported during the lift and the object being lifted will not be damaged by the rigging.

Step 10 Attach the rigging to the load.

Step 11 Signal the crane operator to move the crane hook into position above the load.

Step 12 Arrange the rigging so that no lines are crossed and there are no kinks in the lines, then hang the rigging on the crane hook.

Step 13 Signal the crane operator to slowly raise the hook to take the slack out of the rigging.

Step 14 Check to ensure that the rigging is securely attached to the load and will not slip.

Step 15 Check the hoist line to ensure that it is perfectly vertical.

WARNING! The hoist line must be vertical to the load when the load is lifted or it will swing horizontally and possibly damage the object being lifted as well as other equipment in the area. This also presents a hazardous condition for the rigger and other personnel in the area.

Step 16 Signal the crane operator to slowly test lift the load about 2 inches to ensure that the load is properly rigged to the center of gravity and is lifting level and stable.

Note If the load is not lifting level and stable, signal the crane operator to lower it to the ground. Adjust the rigging as needed.

Step 17 Repeat Steps 15 and 16 until the load is properly rigged and can be moved safely.

Step 18 Check the route that the load will be moved to ensure that no personnel will be under the load at any time and that there are no obstructions, such as overhead wires or cables.

Step 19 Clear the area where the load will be placed, and put down dunnage to set the load on if needed.

Step 20 Attach a tag line to the load if needed to control the load once it is lifted.

Note Some job sites require a tag line on every lift.

CORE CURRICULA TRAINEE TASK MODULE 00106

Step 21 Signal the crane operator to slowly lift the load to a height that will allow it to clear all obstructions.

Note Lifting and moving the load requires a combination of maneuvers by the crane operator. The rigger must know the hand signals to direct these maneuvers.

Step 22 Signal the crane operator to move the load to a position just above the intended location.

Step 23 Arrange the dunnage to set the load on properly.

WARNING! Keep all parts of your body out from under suspended loads to prevent possible serious bodily injury.

Step 24 Signal the crane operator to slowly lower the load, and position it on the dunnage.

Step 25 Signal the crane operator to lower the crane hook, and remove the rigging from the hook.

Step 26 Disconnect the rigging from the load.

Step 27 Store all rigging in the proper place.

REVIEW QUESTIONS #8

1. What is considered the load for a crane?
2. What should you do to prevent anyone from getting near the crane while it is in operation?
3. What causes a load to swing when it is lifted?

PERFORMANCE/LABORATORY EXERCISE

1. Practice rigging and moving materials and equipment as directed by your instructor.

SUMMARY

Safety is the first concern when performing rigging operations. Rigging equipment comes in many shapes and sizes for different rigging jobs. Selecting the proper rigging for the job is essential. Rigging equipment should be inspected before every use and at regular intervals during extended use. The crane operator and the rigger must agree on hand signals to be used before starting a rigging job. You should always know the weight and center of gravity of an object before rigging it to ensure that the rigging is not overloaded and the object can be safely lifted.

References

For advanced study of topics covered in this Task Module, the following work is suggested:

Riggers Handbook, Broderick and Bascom Rope Co; 10440 Trenton Ave. St. Louis, Mo 63132

Answers to Review Questions

Review Questions #1
1. Core, strand, wire
2. Six
3. No
4. Eye splices
5. Shackles
6. Crane hook
7. Three
8. Choker hitch
9. Sun light damages
10. Choker sling

Review Questions #2
1. Broken wires
2. No
3. Each time used

Review Questions #3
1. To communicate over noise and distance
2. Radio
3. No
4. Raise boom

Review Questions #4
1. See chart/tables
2. 32 inches
3. Calculation

Review Questions #5
1. Bowline
2. Timber hitch
3. Barrel hitch

Review Questions #6
1. A-frame derrick
2. Gin pole derrick
3. Guyed derrick
4. Stiff leg derricks

Review Questions #7
1. Crawler cranes
2. Truck-mounted cranes
3. Truck-mounted hydraulic cranes
4. Gantry-mounted cranes
5. Tower-mounted cranes
6. Hammerhead crane
7. Drotts

Review Questions #8
1. Everything that is suspended by the boom
2. Clear area, check radius
3. Load not rigged to center of gravity